Universes are Particles

Stephen Blaha Ph. D.
Blaha Research

Megaverses are also Particles
QED-like Universe Expansion
Shape of Vacuum Polarization = Shape of Universe Expansion
QED Fine Structure Constant Calculated *Exactly*!
A Universe Fine Structure Constant Determined
Standard Model Coupling Constants Calculated
Phenomenological Scale Factor & Hubble Parameter at t = 0

**A 33 % "BIG DIP" in Universe Expansion at the
Transition to Matter-Dominated**

The Origin of a Universe at an Essential Singularity
Universes Begin with the Same Initial Conditions
Life Prospects are Similar in All Universes at All Times

Pingree-Hill Publishing
MMXXI

Rev. 00/00/01   May 28, 2021

To Margaret

Some Other Books by Stephen Blaha

All the Megaverse! Starships Exploring the Endless Universes of the Cosmos using the Baryonic Force (Blaha Research, Auburn, NH, 2014)

SuperCivilizations: Civilizations as Superorganisms (McMann-Fisher Publishing, Auburn, NH, 2010)

All the Universe! Faster Than Light Tachyon Quark Starships & Particle Accelerators with the LHC as a Prototype Starship Drive Scientific Edition (Pingree-Hill Publishing, Auburn, NH, 2011).

Unification of God Theory and Unified SuperStandard Model THIRD EDITION (Pingree Hill Publishing, Auburn, NH, 2018).

The Exact QED Calculation of the Fine Structure Constant Implies ALL 4D Universes have the Same Physics/Life Prospects (Pingree Hill Publishing, Auburn, NH, 2019).

Unified SuperStandard Theory and the SuperUniverse Model: The Foundation of Science (Pingree Hill Publishing, Auburn, NH, 2018).

Quaternion Unified SuperStandard Theory (The QUeST) and Megaverse Octonion SuperStandard Theory (MOST) (Pingree Hill Publishing, Auburn, NH, 2020).

Unified SuperStandard Theories for Quaternion Universes & The Octonion Megaverse (Pingree Hill Publishing, Auburn, NH, 2020).

The Essence of Eternity: Quaternion & Octonion SuperStandard Theories (Pingree Hill Publishing, Auburn, NH, 2020).

A Very Conscious Universe (Pingree Hill Publishing, Auburn, NH, 2020).

Beyond Octonion Cosmology (Pingree Hill Publishing, Auburn, NH, 2021).

Available on Amazon.com, bn.com, Amazon.co.uk and other international web sites as well as at better bookstores (through Ingram Distributors).

CONTENTS

FIGURES and TABLES

Introduction

Elementary particles are very small and of "fixed" size. Universes (at least the one we know) are very large and expand from the very small. In this book we present evidence for a particle interpretation of universes. They originate from a size comparable to elementary particles and expand, as we show, to enormous size in a manner that approximates the form of the vacuum polarization of an elementary particle such as an electron.

The vacuum polarization of a charged particle was calculated by this author to great approximation to give an exact value (as experimentally known to 14 places) for the Fine Structure Constant α. The vacuum polarization, when transformed to a time dependent form, corresponds to a major part of the scale factor $a(t)$ for an expanding universe as determined by Hubble Parameter measurements. This remarkable similarity combined with a factor embodying an essential singularity at the origin of the universe, which has also been conjectured for QED vacuum polarization, provides an excellent phenomenological description of a universe. In this view a universe is generated by fermion-antifermion annihilation in a higher space such as we see in Octonion Cosmology. The scale factor marks this point with an essential singularity. The Big Bang then starts evolution with the same form as vacuum polarization.

The success of the author's approach to vacuum polarization is reflected in its ability to also approximately calculate the coupling constants for the ElectroWeak and Strong Interactions.

Some of the high points described in the book are:

1. Megaverses are also Particles
2. Shape of Vacuum Polarization = Shape of Universe Expansion
3. QED Fine Structure Constant Calculated *Exactly*!
4. A Universe Fine Structure Constant Determined
5. Standard Model Coupling Constants Calculated
6. Phenomenological Scale Factor & Hubble Parameter at $t = 0$
7. A 33 % "BIG DIP" in Universe Expansion at the Transition to Matter-Dominated
8. The Origin of a Universe at an Essential Singularity
9. Universes Begin with the Same Initial Conditions
10. Life Prospects are Similar in All Universes at All Times

The calculation of the vacuum polarization and α, which are strictly local quantum field theory calculations, shows that α is not time dependent, and is independent of location. Thus in our universe, and any other universes, α is the same. Universe evolution is based on an unchanging α. Chemistry and Biology are the "same" everywhere. The success of the calculations has profound consequences.

The calculation results (from chapter 2) are

$g =$	-0.0005805369 0000	-0.0005805369 1948	-0.0005805369 5000
$\alpha =$	0.007297352	***0.0072973525693***	0.007297353
$F_2 \times 10^{10} =$	3.26316 <u>0681</u>7671	3.26316 <u>025</u>452474	3.26316 <u>134</u>861337

Table 2.1. Values of g, α and $F_2(\alpha) \times 10^{10}$. F_2 is very close to zero for the displayed range of values and throughout the flat region. F_2 has a local **minimum** at precisely the known value of $\alpha = 0.0072973525693$ (11).

1. Space Instances as Particles in Octonion Cosmology

There is experimental and theoretical evidence for viewing the universe as within a larger space.[1] The details of the universe then become interesting in a number of respects. How is it expanding? What drives the universe expansion? What is the nature of the universe expansion? What lies beyond the universe boundary? Do new forces appear at the boundary of the universe? And so on.

In the following chapters we present a theory with answers to some of these questions. We also generalize the discussion to Megaverses and to the other spaces of Octonion Cosmology.

Our findings are:

1. There is experimental and theoretical evidence for entities beyond the boundaries of the universe. See Blaha (2021c) for details.

2. The expansion of our universe (and possibly other universes and Megaverses) appears to be very similar (even quantitavely) to the QED vacuum polarization of particles.

3. It is possible that our universe is a Black Hole.

We propose to begin by exploring the connection between QED-like vacuum polarization and the expansion of universes.

---

[1] Blaha (2021c).

2. Vacuum Polarization in QED and the Fine Structure Constant α

In Blaha (2019f) we calculated the value of the QED Fine Structure Constant α = 0.0072973525693 *exactly*. The calculation was independent of any external factors. The successful determination of α is due solely to vacuum polarization in QED. The success of the calculation has major implications:

1. The value is independent of time. Consequently it has been the same since the universe was created.

2. The value is locally determined. It is independent of location in the universe.

3. The extreme locality of the vacuum polarization, and α, makes them independent of space-time curvature, and matter and energy distributions.

4. All models of the universe *must* use the known value of α. It cannot be specified differently.

5. All valid experimental data must be consistent with the known value of α.

6. All studies of life and matter in all environments must use the known value of α. This applies particularly to evolution.

2.1 The Calculation of α

The calculation of α is presented in Appendices 2-A and 2-B.

We have examined the values of the quantities in eq. 2-A.8 looking for an essential singularity (eq. 2-A.2) or its approximation. Fig. 2.1 below plots $F_2(\alpha)$ as a function of g. It displays a "flat region." While essential singularities usually are thought to imply a transcendental function such as $\exp(1/\alpha)$, a constant function with value zero fulfills the essential singularity conditions in eq. 2-A.1. Therefore we take the "flat region" to indicate an essential singularity.

Fig.[2] 2.1 shows a "close up" of the flat region where F_2 is approximately zero. Upon close numeric analysis we find the results in Tables 2.1 and 2.2.

---

[2] These figures appeared in Blaha (2019a) and (2019b).

$g =$	-0.0005805369 0000	-0.0005805369 1948	-0.0005805369 5000
$\alpha =$	0.007297352	*0.0072973525693*	0.007297353
$F_2 \times 10^{10} =$	3.26316 06817671	3.26316 025452474	3.26316 134861337

Table 2.1. Values of g, α and $F_2(\alpha) \times 10^{10}$. F_2 is very close to zero for the displayed range of values and throughout the flat region. F_2 has a local **minimum** at precisely the known value of α = 0.0072973525693 (11).

$g =$	-0.00058053700	-0.00058053705	-0.00058053710
$\alpha =$	0.007297354	0.007297354	0.007297355
$F_2 \times 10^{10} =$	3.26316 299072544	3.26316 29663526	3.26316 408259273

Table 2.2. Other neighboring values of g, α and $F_2(\alpha) \times 10^{10}$ in the flat region *away* from g = 0 (where our approximate F_2 is exactly zero.) F_2 is very close to zero for the displayed range of values and throughout the flat region.

Thus we have a very good approximation with $F_2(\alpha) \cong 0$ at the experimentally known value that is exact to 13 places with a minimum in $F_2(\alpha)$ as anticipated.

F_2 is nearly zero, as are its derivatives, at the physical Fine Structure Constant. It closely approximates a trivial essential singularity of constant value zero in a neighborhood of the singularity.

Note $F_2(\alpha = 0) = 0$ as well. This zero can be viewed as a type of singularity. If QED could transition from positive α to negative α then it would lead to a catastrophe since like charges would then attract.[3,4]

It is extremely important to note the calculation is strictly QED. Thus α is space and time independent, and not Anthropic.

---

[3] Freeman Dyson has speculated on this possibility.

[4] $F_2(\alpha)$ may have more than one zero. One of the zeroes is at the value of the Fine Structure Constant as we show.

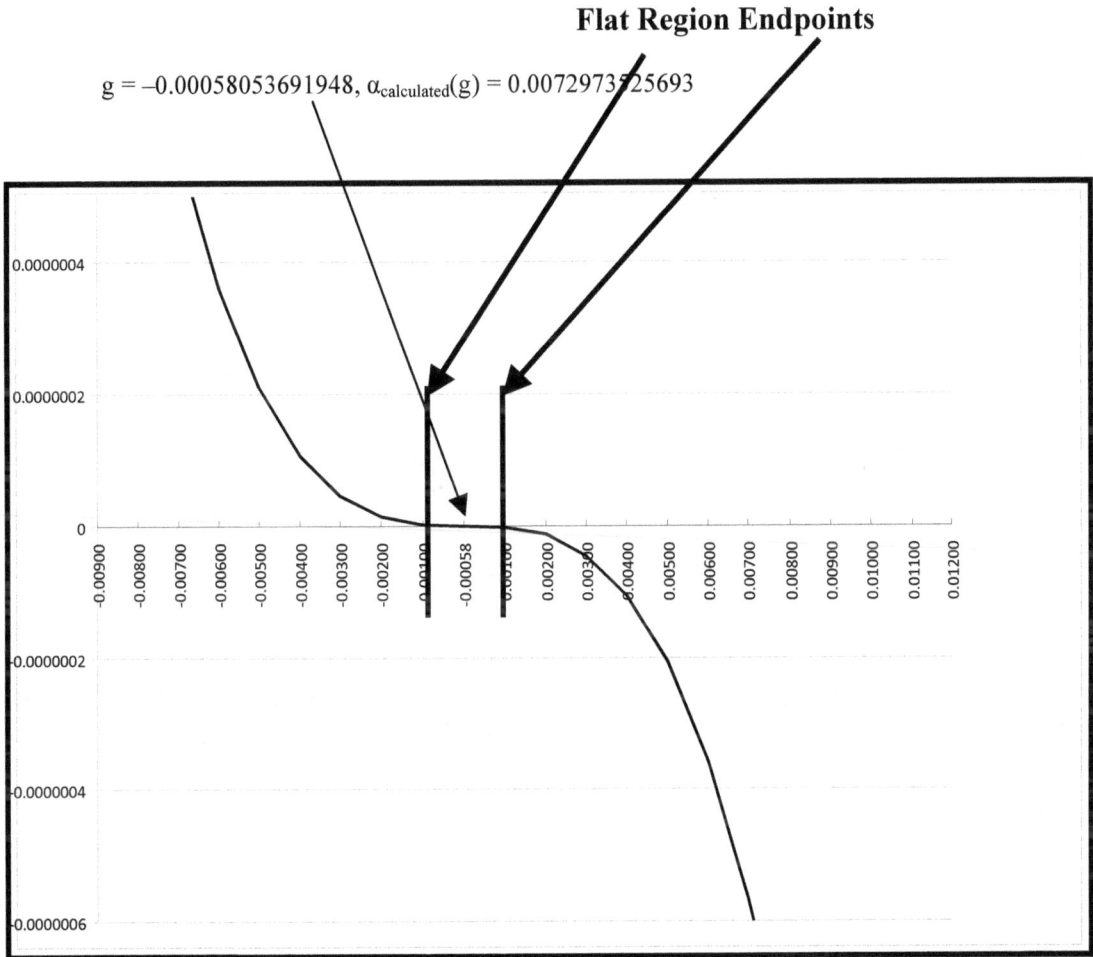

Figure 2.1. Close up plot of our eigenvalue function $F_2(g)$ (vertical axis) vs. g.

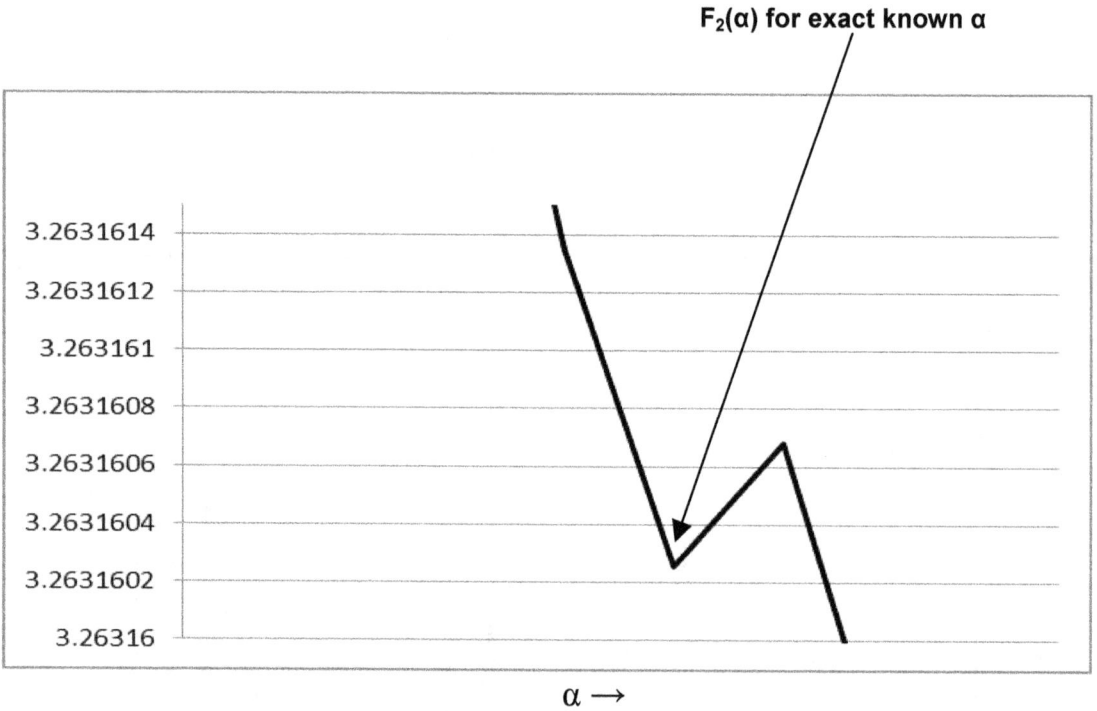

Figure 2.2. Detailed closeup plot of $F_2(\alpha) \times 10^{10}$ data in Tables 2.1 and 2.2. The local minimum of $F_2(\alpha) \times 10^{10}$ at g = -0.00058053691948 corresponds to the exact known value of $\alpha = 0.0072973525693$.

Appendix 2-A. The Solution for the Johnson-Baker-Willey Eigenvalue Function

In a series of remarkable papers Johnson, Baker and Willey[5] developed a finite theory of massless QED (called JBW) without divergences if a certain function $F_1(\alpha)$ of the fine structure constant α called the eigenvalue function were zero. (A zero would imply Z_3, the divergent vacuum polarization constant of the electron, was zero.)

Adler[6] refined the discussion by pointing out that a zero of the eigenvalue function might be an essential singularity with:

$$F_1(\alpha) = 0 \qquad\qquad (2\text{-A.1})$$
$$d^n F_1(\alpha)/d\alpha^n = 0$$

The calculation of the eigenvalue function was reduced by JBW to the sum of all single loop vacuum polarization diagrams of the general form of Fig. 2-A.1.

In 1973 the author[7] calculated $F_1(\alpha)$ approximately to all orders in α. A search for an essential singularity proved fruitless. Recently the author noticed that the vacuum polarization of the electron is manifest in experiment with the effective value of α increasing at higher energies. Thus Z_3 is not zero and has a divergent piece.

On this basis the author proposed, in a series of books in 2019 that the *appropriate* eigenvalue condition was

$$F_2(\alpha) = 0$$

where

$$F_2(\alpha) = F_1(\alpha) - [2/3 + \alpha/(2\pi) - (1/4)[\alpha/(2\pi)]^2] \qquad (2\text{-A.2})$$

The additional terms are those appearing in the exact low order calculation of $F_1(x)$:

$$F_{1\text{ low order}}(\alpha) = 2/3 + \alpha/(2\pi) - (1/4)[\alpha/(2\pi)]^2 \qquad (2\text{-A.3})$$

In terms of F_2 the renormalization constant Z_3 is

$$Z_3 = 1 + F_1(\alpha)\ln(p/\Lambda) = 1 + F_2(\alpha) + \text{divergent terms} = 1 + \text{divergent terms} \qquad (2\text{-A.4})$$

The original goal of the JBW Model was to solve massless QED in a manner that made all renormalization constants either 1 or at least finite.

---

[5] Summarized in some detail in K. Johnson and M. Baker, Phys. Rev. **D8**, 1110 (1973). Also in Blaha (2019b) and (2019c).
[6] S. Adler, Phys. Rev. **D5**, 3021 (1972).
[7] S. Blaha, Phys. Rev. **D9**, 2246 (1974). See Appendix 2-B.

We modified this goal. We shall see that we can obtain a physically improved eigenvalue function F_2 that has a zero at the known fine structure constant α. Until now we have not specified the value α that appears in the preceding equations. We now define α as a partially renormalized quantity that is related to the bare fine structure constant α_0 by

$$\alpha = \alpha_0[2/3 + \alpha_0/(2\pi) - (1/4)[\alpha_0/(2\pi)]^2] \qquad (2\text{-A}.5)$$

We show in chapter 2 that the evaluation of the F_2 eigenvalue function gives the known approximate[8] physical value[9] of the fine structure constant:

$$\alpha = 0.0072973525693 \ (11) \qquad (2\text{-A}.6)$$

The renormalized expressions appearing below are not fully finite. However the intermediate renormalized finite α is physically sensible—more so than the completely finite renormalization constants goal of the JBW Model.

The bare charge constant α_0 is known to approach ∞ at very short distances. The simplest examples of this phenomenon are the physical Coulomb scattering amplitudes and the first order change in hydrogen-like atomic energy levels.[10] Thus our modified JBW Model with a partial renormalization conforms to physical reality:

$$
\begin{aligned}
Z_3 &= 1 + \{\alpha F_2(\alpha) + \alpha[2/3 + \alpha/(2\pi) - (1/4)[\alpha/(2\pi)]^2]\}\ln(p/\Lambda) \\
&= 1 + \alpha\{2/3 + \alpha/(2\pi) - (1/4)[\alpha/(2\pi)]^2\}\ln(p/\Lambda)
\end{aligned}
\qquad (2\text{-A}.7)
$$

at α = the physical Fine Structure Constant where $F_2(\alpha) = 0$.

Our approximate 1973 solution, which summed one loop pieces of the vacuum polarization yielded the algebraic equations, is:[11]

$$A_1 = (g + 1)(1 - 2g^2)/[(g + 2)(g - 1)] \qquad (2\text{-A}.8)$$

$$A_2 = [8g^2(2g + 1) - (2g^3 + 2g^2 + g - 2)(g^2 + 2g + 2)]/[2(g^2 - 1)(g^2 - 4)]$$

$$A_3 = -2(1 + 3g + 6g^2 + 2g^3)/[g(g + 1)]$$

$$A_4 = -(g + 2)(1 + 5g + 6g^2 + 2g^3)/[g(g^2 - 1)] - 1/(g + 1)$$

$$\psi = [gA_3 - (4 + 2g)A_1]/[(4 + 2g)A_2 - g\,A_4]$$

$$(\alpha/2\pi) = [gA_4 - (4 + 2g)A_2]/(A_4A_1 - A_2A_3)$$

---

[8] The constant α is an irrational number.
[9] 2018 CODATA: P. J. Mohr *et al* CODATA group (2019)
[10] See E. A. Ueling, Phys. Rev. **48**, 55 (1935) and R. Serber, Phys. Rev. **48**, 49 (1935).
[11] Blaha *op. cit.*

$F_1(g) = (2/3)(1 - 3g^2/2 - g^3) - (\alpha/4\pi)[(2 + 4g + 4g^2)(g - 2) + \alpha\psi g^3]/[(g^2 - 1)(g - 2) + \alpha(2 + 4g + 4g^2)(g - 2) + \alpha\psi g^3]$

expressed as a function[12] of g (the power of the divergent factor p/Λ) with ψ specifying the gauge, and with the renormalization definitions

$$\Gamma_\mu(p) = f(\gamma_\mu + 2g\gamma \cdot pp_\mu/p^2)(p/\Lambda)^{2g} \tag{2-A.9}$$
$$S_F = [f\gamma \cdot p(p/\Lambda)^{2g}]^{-1} \tag{2-A.10}$$
$$\Gamma_{\mu\alpha}(p) = (f_3/p^2)(\gamma \cdot p\gamma_\mu\gamma_\alpha - \gamma_\alpha\gamma_\mu\gamma \cdot p)(p/\Lambda)^{2g} \tag{2-A.11}$$

and

$$F_1 = (2/3)(1 - 3g^2/2 - g^3) - f_3/f \tag{2-A.12}$$

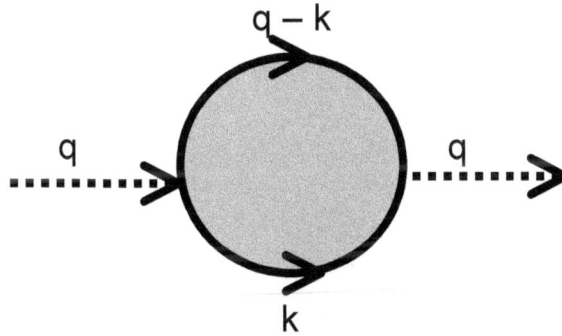

Figure 2-A.1 One loop vacuum polarization Feynman diagrams with internal free photon propagators.

We thus have an expression for the eigenvalue function F_2 within the framework of massless QED. We shall see that the eigenvalue function generalizes to all Standard Model gauge interactions. We shall also see that it applies to the expansion and contractions of universes upon the introduction of a Dark Energy gauge interaction for universes.

In Blaha (2019b) we generalized eq. 2-A.8 to a "universal" eigenvalue function F_2 to include the Weak interaction and the Strong interaction coupling constants by inserting an interaction specific factor in the α_G equation:

$$(\alpha_G/2\pi) = c_G^{-1}[gA_4 - (4 + 2g)A_2]/(A_4A_1 - A_2A_3) \tag{2-A.13}$$

For a non-abelian group we set

$$c_G^{-1} = [(11/3)C_{ad} - 2C_f/3]/(16\pi)^3 \tag{2-A.14}$$

---

[12] The solution for the eigenvalue function is clearly best expressed in terms of the g factor in the exponents of the divergent renormalization factors. We use $F_1(g)$ and $F_1(\alpha(g))$ interchangeably.

where C_{ad} is the dimension of the fundamental representation of the group and C_f is the number of fermions (fermion flavor) of the interaction.

In chapter 2 above we saw that we obtain the exact QED Fine Structure Constant to the known 13 place accuracy with $F_2 \cong 0$. Thus our "approximate" eigenvalue condition $F_2(\alpha) = 0$ appears to be remarkably accurate. It generalizes directly in chapter 7 to the Standard Model interactions and the universe scale factor.

Blaha (2019b) describes the universal eigenvalue function F_2 in detail including numerous plots.

Appendix 2-B. Johnson-Baker-Willey Eigenvalue Function Paper

In 1974 the author[13] calculated the Johnson-Baker-Willey eigenvalue function for α. The calculation has been verified by a number of independent physicists and found to be correct. This appendix contains the Physical Review article with thanks to the American Physical Society for copyright permission.

---

[13] S. Blaha, Phys. Rev. **D9**, 2246 (1974).

PHYSICAL REVIEW D   VOLUME 9, NUMBER 8   15 APRIL 1974

Approximate calculation of the eigenvalue function in massless quantum electrodynamics*

Stephen Blaha†

Department of Physics, University of Washington, Seattle, Washington 98195
(Received 24 August 1973)

We solve a vertex equation in massless quantum electrodynamics and use the results to calculate an approximation to the eigenvalue function, F_1, in the Johnson-Baker-Willey model. This approximation consists of a summation of the contributions to F_1 of all one-electron-loop diagrams in which no internal photon lines intersect (if all such lines are drawn within the electron loop). Our result reproduces the known low-order terms, $F_1 = 2/3 + \alpha/2\pi - \frac{1}{4}(\alpha/2\pi)^2$ exactly. In addition we find branch-point singularities and zeros at points corresponding to values for the fine-structure constant of order unity. Nonperturbative solutions of the vertex equation and F_1 are also shown to exist.

I. INTRODUCTION

The apparently divergent renormalization constants in quantum electrodynamics (QED) have been a source of uneasiness for many years. Some have taken this property to reflect a fundamental incompleteness of QED. Johnson, Baker, and Willey[1] have responded to the problem by developing a model QED which has finite renormalization constants if a certain function, denoted F_1, of the fine-structure constant is zero. $F_1(\alpha)$ is the coefficient of $(\alpha/2\pi)\ln\Lambda^2$ in the sum of the contributions of all one-electron-loop vacuum-polarization diagrams to Z_3^{-1}. [Examples of $O(\alpha^2)$ and $O(\alpha^3)$ diagrams contributing to F_1 are given in Figs. 1 and 2.]

Adler[2] enhanced interest in F_1 by noting that any zero of F_1 must be of infinite order (an essential singularity) and by raising the possibility that the zero might occur at the value of the physical fine-structure constant. In this case, the requirement that QED be finite would determine the fine-structure constant, and QED would then be a self-contained theory (but for the choice of the electron mass scale).

Because of the importance of this possibility, a calculation of F_1 would be of great interest. Since an exact calculation appears unlikely at the moment, we have calculated an approximate expression for F_1 based on an extrapolation of a property of its low-order terms. In low order[3]

$$F_1 = \frac{2}{3} + \frac{\alpha}{2\pi} - \frac{1}{4}\left(\frac{\alpha}{2\pi}\right)^2 + \cdots. \qquad (1)$$

The third term was first calculated by Rosner in the Landau gauge. Brandt[4] recalculated this term in the Feynman gauge and found that the contribution to F_1 of a certain subset of the diagrams

summed to zero [see Fig. 1(a)]. These same diagrams contained the only appearances of the zeta function, $\zeta(3)$. The diagrams of Fig. 1(b) sum to the total $O(\alpha^2)$ term of F_1. [Due to Brandt's renormalization procedure, each electromagnetic vertex in the contributing set of Fig. 1(b) should be understood to represent a zero-momentum-transfer vertex function rather than a Dirac matrix.] If Brandt's result generalizes to higher order, one could hope to find F_1 by summing a much simpler subset of diagrams. The obvious generalization is to sum only those diagrams in which no internal photon lines intersect when all photon lines are drawn within the electron loop. Figure 2 shows these diagrams in $O(\alpha^3)$.

By choosing to sum only a special subset of the diagrams, we have lost gauge invariance. However, there are two situations where this might be an acceptable loss: (1) where the sum of the selected subset of diagrams indeed reproduces the exact gauge-invariant result (making the question academic), and (2) where the sum of the diagrams gives the dominant contribution to F_1 so that one can observe the important features, e.g., an essential singularity.

We will calculate the contributions of our selected subset of diagrams to F_1 by finding the vertex function and electron propagator which generate the contributions of these diagrams when substituted in the photon self-energy. The gauge in which we work will be determined by requiring that the vertex function and electron propagator (for the set of diagrams considered) satisfy the differential Ward identity.

In Sec. II we derive the necessary vertex equation. Section III gives the details of the calculation. Section IV contains a discussion of the important features of our results.

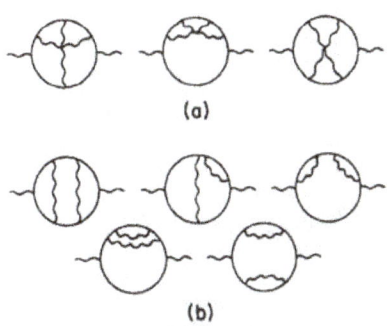

(a)

(b)

FIG. 1. (a) Diagrams whose contribution to F_1 in the Feynman gauge sums to zero. (b) Diagrams which give the total contribution to F_1 in $O(\alpha^2)$ in the Feynman gauge.

II. DERIVATION OF THE VERTEX EQUATION

Before deriving the vertex equation, we will define the relation of F_1 to the vertex function. Johnson, Willey, and Baker[5] relate F_1 and the contribution of one-electron-loop diagrams to $\Pi_{\mu\nu}$, the photon self-energy:

$$\Pi_{\mu\mu\beta\beta} \equiv \frac{\partial^2}{\partial q^\beta \partial q_\beta} \Pi_\mu^\mu(q)\Big|_{q=0}$$

$$= \frac{-12\alpha}{\pi} F_1 \int \frac{dp^2}{p^2} + \cdots, \qquad (2)$$

where q is the external photon momentum. They then show

$$\Pi_{\mu\mu\beta\beta} = \frac{-ie^2}{(2\pi)^4} \int d^4p \, \mathrm{Tr}(\Gamma_\mu G_\beta^\beta \Gamma^\mu + 2\Gamma_{\mu\beta} G^\beta \Gamma^\mu$$

$$+ \Gamma_\mu G K_\beta^\beta G \Gamma^\mu + 2\Gamma_\mu G_\beta K^\beta G \Gamma^\mu$$

$$+ 2\Gamma^\mu G K^\beta G \Gamma_{\mu\beta}), \qquad (3)$$

where G signifies the introduction of the unrenormalized electron propagator S_F at appropriate places in the traces, $S_F \cdots S_F$; K is the electron-positron scattering kernel; Γ_μ is the vertex function; and the appearance of the subscript (superscript) β on a quantity indicates that a derivative is to be taken with respect to the external photon momentum, q, before setting $q=0$. In the case we will consider below K is independent of q: ($K_\beta = K_\beta^\beta = 0$), and the explicit form $\Pi_{\mu\mu\beta\beta}$ assumes is

$$\Pi_{\mu\mu\beta\beta} = \frac{-ie^2}{(2\pi)^4} \mathrm{Tr} \int d^4p \, (\Gamma_\mu S_F \Gamma^\mu S_{F\beta}^\beta + \Gamma_\mu S_{F\beta}^\beta \Gamma^\mu S_F$$

$$+ 2\Gamma^\mu S_{F\beta} \Gamma_\mu S_F^\beta + 2\Gamma_\mu S_F^\beta \Gamma_\beta^\mu S_F$$

$$+ 2\Gamma^\mu S_F \Gamma_{\mu\beta} S_F^\beta). \qquad (4)$$

The quantities we must find are $\Gamma_\mu(p,p)$, the zero-momentum-transfer vertex function, S_F, and $\Gamma_{\mu\beta} = (\partial/\partial q^\beta)\Gamma_\mu|_{q=0}$. We will find expressions for

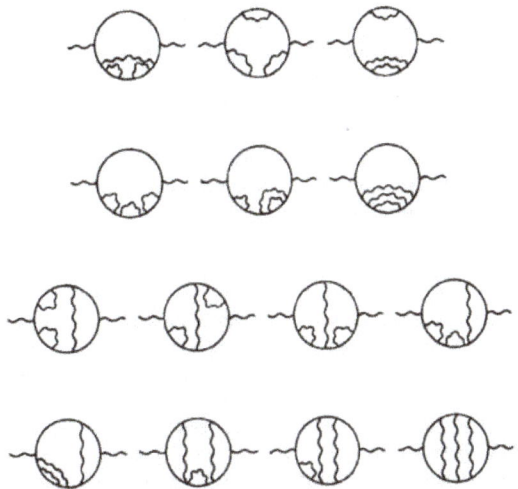

FIG. 2. The distinct diagrams contributing to the $O(\alpha^3)$ term in F_1 in our approximation.

these quantities which generate the contributions to F_1 of the diagrams described in the Introduction.

Our starting point will be the equation for the vertex function in massless QED with no self-energy insertions in the photon propagator:

$$\Gamma_\mu(p_+, p_-) = \gamma_\mu + \frac{ie^2}{(2\pi)^4} \int d^4k \, \Gamma_\lambda(p_+, k_+) S_F(k_+) \Gamma_\mu(k_+, k_-)$$

$$\times S_F(k_-) \Gamma_\sigma(k_-, p_-) D^{\lambda\sigma}(p-k)$$

$$+ \cdots, \qquad (5)$$

where $p_\pm = p \pm \frac{1}{2}q$ are the external electron momenta, $k_\pm = k \pm \frac{1}{2}q$, and $D_{\lambda\sigma}$ is the photon propagator.

Since we are only interested in diagrams with nonintersecting lines, we need only keep the exhibited terms on the right-hand side of the equation. In addition, if we kept strictly to nonintersecting photon line diagrams, we would substitute γ_ν for Γ_ν and γ_σ for Γ_σ in the second term. But the suggestion of Brandt's work was to place zero-momentum-transfer vertex functions at all photon vertices. We choose to do this in the following manner:

$$\Gamma_\mu(p_+, p_-) = \gamma_\mu + \frac{ie^2}{(2\pi)^4} \int d^4k \, \Gamma_\lambda(k,k) S_F(k_+) \Gamma_\mu(k_+, k_-)$$

$$\times S_F(k_-) \Gamma_\sigma(k,k) D^{\lambda\sigma}(p-k), \qquad (6)$$

where we have introduced zero-transfer vertices which are only functions of the loop integration variable. We will discuss the results of other possible choices such as $\Gamma_\nu(p,p) \cdots \Gamma_\sigma(p,p)$ in Sec. IV. Since $\Gamma_\nu(p_+, k_+)$ and $\Gamma_\sigma(k_-, p_-)$ can be expanded in a double power series whose first terms are $\Gamma_\nu(k, k)$ and $\Gamma_\sigma(k, k)$, respectively, it is clear that we are not introducing any spurious contributions by our choice at these vertices. Figure 3 gives a graphical representation of Eq. (6).

Having chosen our vertex equation, we now complete our set of equations by requiring that the Ward identity be satisfied,

$$\Gamma_\mu = \frac{\partial}{\partial p^\mu} S_F^{-1}(p). \tag{7}$$

Equations (6) and (7) can only be simultaneously satisfied in a special gauge. This gauge is the Feynman gauge up to terms of $O(\alpha)$. In a preliminary study[6] of these equations, the gauge function was set equal to zero (which was reasonable in an investigation of the region near $\alpha = 0$) and only the component of the vertex equation kept which could be obtained by multiplying Eq. (6) by γ_μ and summing over μ. The results of that investigation will be seen to be in qualitative agree-

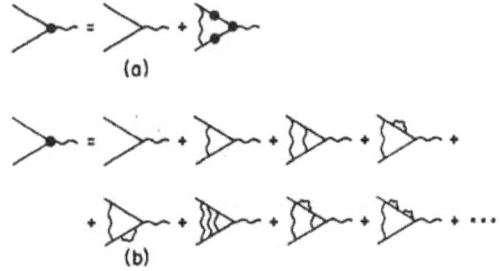

(a)

(b)

FIG. 3. (a) Diagrammatic representation of the vertex equation. (b) Some low-order vertex diagrams embodied in the vertex equation. All seemingly bare vertices (except the vertex to which the external photon couples) actually represent zero-momentum-transfer vertex functions.

ment with these results. To find F_1 we must solve two equations obtained from Eq. (6) by setting $q = 0$ in the first case, and setting $q = 0$ after taking a derivative in the second:

$$\Gamma_\mu(p) = \gamma_\mu + \frac{ie^2}{(2\pi)^4} \int d^4k \, \Gamma_\lambda(k) S_F(k) \Gamma_\mu(k) S_F(k) \Gamma_\sigma(k) D^{\lambda\sigma}(p - k), \tag{8}$$

$$\Gamma_{\mu\alpha} = \frac{ie^2}{(2\pi)^4} \int d^4k \, D^{\lambda\sigma}(p - k) \left[\Gamma_\lambda(k) S_F(k) \Gamma_{\mu\alpha}(k) S_F(k) \Gamma_\sigma(k) \right.$$
$$\left. + \Gamma_\lambda(k) \left(\frac{\partial}{\partial k^\alpha} S_F \right) \Gamma_\mu(k) S_F(k) \Gamma_\sigma(k) - \Gamma_\lambda(k) S_F(k) \Gamma_\mu(k) \left(\frac{\partial}{\partial k^\alpha} S_F(k) \right) \Gamma_\sigma(k) \right], \tag{9}$$

where

$$\Gamma_\mu(k) = \Gamma_\mu(k, k)$$

and

$$\Gamma_{\mu\alpha}(p) = 2 \frac{\partial}{\partial q^\alpha} \Gamma_\mu(p_+, p_-) \Big|_{q=0}, \tag{10}$$

with $p_\pm = p \pm \frac{1}{2}q$. [Note that $\Gamma_{\mu\beta}$ in Eq. (10) is a factor of 2 different from the similar-looking quantity in Eqs. (3) and (4). All subsequent appearances of $\Gamma_{\mu\beta}$ conform to Eq. (10).] The photon propagator has the form

$$D_{\mu\nu}(k) = -\left(\frac{g_{\mu\nu}}{k^2} - \psi \frac{k_\mu k_\nu}{k^4} \right), \tag{11}$$

where $\psi = \psi(\alpha)$ will be determined by requiring Eqs. (7) and (8) to be consistent. In the following we will perform a Wick rotation on all four-vectors in Eqs. (7), (8), and (9), and introduce an ultraviolet cutoff, Λ, in all radial integrations.

One approach to solving our equations would be to iterate Eq. (8) and calculate Γ_μ and S_F order by order in α. The first iteration of Eq. (8) gives the triangle diagram which contributes surface terms to Γ_μ which are proportional to p^2/Λ^2. For

terms of this sort we follow the procedure of Gell-Mann and Low[7] (Sec. III) and "drop terms that approach zero as $\Lambda^2 \to \infty$." Order by order in α such terms should be dropped as they appear. As a result

$$S_F^{-1} \equiv \not{p} B(p^2) = \sum_{i,j=0}^{\infty} c_{ij} \alpha^i [\ln(p^2/\Lambda^2)]^j,$$

with c_{ij} a constant satisfying $c_{ij} = 0$ if $j > j_0(i)$; i.e., in any order of α there is an upper bound on the power of divergent logarithms. Equations (6) and (7) allow us to calculate c_{ij} recursively (dropping terms violating the criteria on c_{ij} such as p^2/Λ^2).

Actually, since the terms we are dropping are surface terms we may find the form of the "kept" terms of B by converting Eq. (6) into a differential equation after performing angular integrations. The surface terms are now eliminated and scaling arguments show $B = f(p/\Lambda)^{2g}$ with f and g constants. This is not a solution of Eq. (6), because of surface terms. However, substitution of this solution into the right-hand side of Eq. (6) and expansion of terms in double power series

$$\sum d_{ij} \alpha^i [\ln(p^2/\Lambda^2)]^j$$

will allow us to uniquely determine the surface terms which would have been dropped in a recursive solution. In the case at hand we drop terms proportional to p^2/Λ^2 so that equating coefficients on either side of the equation will not lead to a violation of $c_{ij}=0$ for $j > j_0(i)$ in B's expansion.

As a result of the form of B we have

$$\Gamma_\mu(p) = f\left(\gamma_\mu + 2g\frac{\slashed{p}p_\mu}{p^2}\right)\left(\frac{p}{\Lambda}\right)^{2\epsilon}, \tag{12}$$

$$S_F = \left[f\slashed{p}\left(\frac{p}{\Lambda}\right)^{2\epsilon}\right]^{-1}, \tag{13}$$

$$\Gamma_{\mu\alpha} = \frac{f_3}{p^2}(\slashed{p}\gamma_\mu\gamma_\alpha - \gamma_\alpha\gamma_\mu\slashed{p})\left(\frac{p}{\Lambda}\right)^{2\epsilon}, \tag{14}$$

which implies

$$F_1 = \tfrac{2}{3}(1 - \tfrac{3}{2}g^2 - g^3) - f_3/f. \tag{15}$$

Equation (9) determines f_3, which is the product of a function of g and ψ times f. Thus f_1 depends only on g and ψ [cf. Eq. (58)] and surface terms of the kind mentioned above do not affect F_1.

Rather than calculate F_1 using the approach just outlined, we will substitute Eqs. (12), (13), and (14) into Eqs. (8) and (9) and determine g, ψ, f, and f_3 in this more economical way.

III. THE CALCULATION

In this section we solve Eqs. (7), (8), and (9) by showing that Eqs. (12), (13), and (14) are consistent solutions of the equations for appropriate choices of f, g, and f_3.

We begin by substituting Eqs. (12) and (13) into Eq. (8). In order to illustrate our approach, we will now use a fact which will be apparent at the end; namely, that Eq. (8) has the following form after performing the loop integral:

$$f\left(\gamma_\mu + 2g\frac{p_\mu\slashed{p}}{p^2}\right)\left(\frac{p}{\Lambda}\right)^{2\epsilon} = \gamma_\mu + A\gamma_\mu + B\gamma_\mu\left(\frac{p}{\Lambda}\right)^{2\epsilon} + C\frac{\slashed{p}p_\mu}{p^2}\left(\frac{p}{\Lambda}\right)^{2\epsilon} + D\frac{\slashed{p}p_\mu}{p^2}, \tag{16}$$

where A, B, C, D are constants. As a result

$$f = B, \tag{17}$$

$$2gf = C, \tag{18}$$

$$1 + A = 0, \tag{19}$$

$$D = 0 \tag{20}$$

are required for consistency. We will begin by calculating the two relations resulting from multiplying Eq. (16) by γ_μ and summing over μ:

$$f(4 + 2g) = 4B + C \tag{21}$$

and

$$0 = 1 + A, \tag{22}$$

assuming $D = 0$. Later we will calculate C and show $D = 0$.

Initially Eq. (8) has the form

$$f(4 + 2g)\left(\frac{p}{\Lambda}\right)^{2\epsilon} = 4 + \frac{ie^2 f}{4(2\pi)^4}\int\frac{d^4k}{k^4}D^{\lambda\sigma}(p-k)\left(\frac{k}{\Lambda}\right)^{2\epsilon}\mathrm{Tr}\gamma^\mu(\gamma_\lambda\slashed{k} + 2gk_\lambda)\left(\gamma_\mu + 2g\frac{\slashed{k}k_\mu}{k^2}\right)(\slashed{k}\gamma_\sigma + 2gk_\sigma) \tag{23}$$

after taking the trace of γ_μ times Eq. (8).

The γ-matrix algebra may be performed in the integrand, and one arrives at the following expression:

$$f(4 + 2g)\left(\frac{p}{\Lambda}\right)^{2\epsilon} = 4 - \frac{ie^2 f}{(2\pi)^4}\int\frac{d^4k\,k^{2\epsilon-2}}{(p-k)^2\Lambda^{2\epsilon}}\left(4w - \frac{\psi}{(p-k)^2k^2}\{4v[k\cdot(p-k)]^2 - 2gk^2(p-k)^2\}\right) \tag{24}$$

or

$$f(4 + 2g)\left(\frac{p}{\Lambda}\right)^{2\epsilon} = 4 - \frac{ie^2 f}{(2\pi)^4}[(4w + 2g\psi)I_1 - \psi v(I_2 + 2I_3 + I_4)], \tag{25}$$

where $w = 1 + 3g + 6g^2 + 2g^3$, $v = 1 + 5g + 6g^2 + 2g^3$, and

$$I_1 = \int\frac{d^4k}{(p-k)^2k^2}\left(\frac{k}{\Lambda}\right)^{2\epsilon} \equiv \frac{i\pi^2}{g}\left[1 - \frac{1}{g+1}\left(\frac{p}{\Lambda}\right)^{2\epsilon}\right], \tag{26}$$

$$I_2 = \int\frac{d^4k}{k^4}\left(\frac{k}{\Lambda}\right)^{2\epsilon} \equiv \frac{i\pi^2}{g}, \tag{27}$$

$$I_3 = \int\frac{d^4k\,k^2(k^2-p^2)}{(p-k)^2k^4}\left(\frac{k}{\Lambda}\right)^{2\epsilon} \equiv \frac{i\pi^2}{g}\left[1 + \frac{2}{g^2-1}\left(\frac{p}{\Lambda}\right)^{2\epsilon}\right], \tag{28}$$

$$I_4 = \int\frac{d^4k(k^2-p^2)^2}{(p-k)^4k^4}\left(\frac{k}{\Lambda}\right)^{2\epsilon} \equiv \frac{i\pi^2}{g}\left[1 + \frac{2g}{g^2-1}\left(\frac{p}{\Lambda}\right)^{2\epsilon}\right]. \tag{29}$$

We have assumed $g > 0$ and used

$$4[k\cdot(p-k)]^2 = (p^2 - k^2)^2 - 2(p^2 - k^2)(p-k)^2 + (p-k)^4.$$

The method of evaluating the integrals given above is described in the Appendix. Substituting for the integrals in Eq. (25), we obtain the following two equations:

$$4 + 2g = \frac{-\alpha}{4\pi}\left[\frac{4w}{g(g+1)} + \frac{2\psi}{g+1} + \frac{2\psi(g+2)v}{g(g^2-1)}\right], \quad (30)$$

$$0 = 4g + \frac{\alpha}{\pi}fw + \frac{\alpha}{2\pi}gf\psi - \frac{\alpha}{\pi}\psi fv. \quad (31)$$

Equations (30) and (31) correspond to Eqs. (21) and (22). We now proceed to calculate C and show $D = 0$. This allows us to simplify the integrand of Eq. (8) by neglecting all terms proportional to γ_μ. Terms appearing on the right-hand side of Eqs. (32)–(39) are to be taken modulo quantities proportional to γ_μ. Thus

$$2gf\frac{\not{p}p_\mu}{p^2}\left(\frac{p}{\Lambda}\right)^{2\epsilon} = \frac{-ie^2f}{(2\pi)^4}\int\frac{d^4k}{k^2(p-k)^2}\left(\frac{k}{\Lambda}\right)^{2\epsilon}\left[\frac{4\not{k}k_\mu}{k^2}(1+g)(2g^2-1) - \frac{\psi}{(p-k)^2k^2}\left(2(k_\mu p^2 - p_\mu k^2)(\not{p}-\not{k}) - 2\not{p}k_\mu(p-k)^2\right.\right.$$
$$+ 2gk_\mu(p^2-k^2)(\not{p}-\not{k}) - 2gk_\mu(p-k)^2\not{p}$$
$$\left.\left. + 8(g^2+g^3)\frac{k_\mu\not{k}}{k^2}[k\cdot(p-k)]^2 + 8g^2k\cdot(p-k)\frac{k_\mu}{k^2}(k^2\not{p}-k\cdot p\not{k})\right)\right] \quad (32)$$

or

$$2g\frac{\not{p}p_\mu}{p^2}\left(\frac{p}{\Lambda}\right)^{2\epsilon} = \frac{-ie^2}{(2\pi)^4}\left[4(1+g)(2g^2-1)J_{0\mu} - 2\psi J_{1\mu} + 2\not{p}\psi J_{2\mu} - 2g\psi J_{3\mu} + 2g\psi\not{p}J_{2\mu} - 8g^2(g+1)J_{4\mu} - 8g^2\psi J_{5\mu}\right], \quad (33)$$

where

$$J_{0\mu} = \int\frac{d^4k\,k_\mu\not{k}}{k^4(p-k)^2}\left(\frac{k}{\Lambda}\right)^{2\epsilon} \equiv \frac{-i\pi^2 p_\mu\not{p}}{p^2(g+2)(g-1)}\left(\frac{p}{\Lambda}\right)^{2\epsilon}, \quad (34)$$

$$J_{1\mu} = \int\frac{d^4k(\not{p}-\not{k})}{(p-k)^4k^4}(p^2k_\mu - p_\mu k^2) \equiv 0, \quad (35)$$

$$J_{2\mu} = \int\frac{d^4k\,k_\mu}{k^4(p-k)^2}\left(\frac{k}{\Lambda}\right)^{2\epsilon} \equiv \frac{-i\pi^2 p_\mu}{(g^2-1)p^2}\left(\frac{p}{\Lambda}\right)^{2\epsilon}, \quad (36)$$

$$J_{3\mu} = \int\frac{d^4k\,k_\mu(p^2-k^2)(\not{p}-\not{k})}{(p-k)^4k^4}\left(\frac{k}{\Lambda}\right)^{2\epsilon} \equiv \frac{i\pi^2 p_\mu\not{p}}{(g+1)(g+2)p^2}\left(\frac{p}{\Lambda}\right)^{2\epsilon}, \quad (37)$$

$$J_{4\mu} = \int\frac{d^4k\,k_\mu\not{k}[k\cdot(p-k)]^2}{k^6(p-k)^4}\left(\frac{k}{\Lambda}\right)^{2\epsilon} \equiv \frac{i\pi^2(g^2+2g+2)p_\mu\not{p}}{2(g^2-1)(g^2-4)p^2}\left(\frac{p}{\Lambda}\right)^{2\epsilon}, \quad (38)$$

$$J_{5\mu} = \int\frac{d^4k\,k_\mu k\cdot(p-k)(k^2\not{p}-k\cdot p\not{k})}{(p-k)^4k^6}\left(\frac{k}{\Lambda}\right)^{2\epsilon} \equiv \frac{-i\pi^2(2g+1)p_\mu\not{p}}{(g^2-1)(g^2-4)p^3}\left(\frac{p}{\Lambda}\right)^{2\epsilon}. \quad (39)$$

The evaluation of these integrals is discussed in the Appendix.
Substitution of the above expressions into Eq. (33) gives

$$g = \frac{\alpha(g+1)(1-2g^2)}{2\pi(g+2)(g-1)} - \frac{\alpha}{4\pi}\psi\frac{[(2g^3+2g^2+g-2)(g^2+2g+2) - 8g^2(2g+1)]}{(g^2-1)(g^2-4)} \quad (40)$$

after some algebra. In addition it is clear that $D = 0$ by examination of Eqs. (34)–(39).

Equations (30), (31), and (40) should now be solved to find $f(\alpha)$, $g(\alpha)$, and $\psi(\alpha)$. However, it is obvious that such an approach would not lead to intelligible results. Therefore, *we will parametrize α, ψ, and f with g*. We easily obtain

$$\frac{\alpha}{2\pi} = \frac{gA_4 - (4+2g)A_2}{A_4 A_1 - A_2 A_3}, \quad (41)$$

where

$$A_1 = \frac{(g+1)(1-2g^2)}{(g+2)(g-1)}, \quad (42)$$

$$A_2 = \frac{-(2g^3+2g^2+g-2)(g^2+2g+2) + 8g^2(2g+1)}{2(g^2-1)(g^2-4)}, \quad (43)$$

$$A_3 = \frac{-2(1+3g+6g^2+2g^3)}{g(g+1)}, \quad (44)$$

$$A_4 = \frac{-(g+2)(1+5g+6g^2+2g^3)}{g(g^2-1)} - \frac{1}{g+1}. \quad (45)$$

In addition

$$\psi = \frac{gA_3 - (4+2g)A_1}{(4+2g)A_2 - gA_4} \quad (46)$$

and

$$f = \frac{-8\pi g}{\alpha[2 + 6g + 12g^2 + 4g^3 - \psi(2 + 9g + 12g^2 + 4g^3)]}. \quad (47)$$

In Sec. IV we will discuss the properties of the above expressions and investigate the behavior of the above quantities in low order of α.

We now proceed to solve Eq. (9) for $\Gamma_{\mu\alpha}$. Substitution of Eqs. (12), (13), and (14) into Eq. (9) leads to

$$\Gamma_{\mu\alpha}(p) = \frac{-ie^2}{(2\pi)^4} \int \frac{d^4k(f_3 - f)}{k^4(p-k)^2} \left(\frac{k}{\Lambda}\right)^{2\epsilon}$$

$$\times \left[(2 + 4g + 4g^2)(\not k \gamma_\mu \gamma_\alpha - \gamma_\alpha \gamma_\mu \not k) \right.$$

$$- \frac{\psi}{k^2(p-k)^2} (\{k^4 + k^2 p^2 - 4gk^2 k \cdot (p-k) + 4g^2[k \cdot (p-k)]^2\}(\not k \gamma_\mu \gamma_\alpha - \gamma_\alpha \gamma_\mu \not k)$$

$$\left. + [-k^2 + 2gk \cdot (p-k)] [\not p \not k (\not k \gamma_\mu \gamma_\alpha - \gamma_\alpha \gamma_\mu \not k) + (\not k \gamma_\mu \gamma_\alpha - \gamma_\alpha \gamma_\mu \not k)\not k \not p]) \right]. \quad (48)$$

We multiply Eq. (48) by $\not p \gamma_\alpha \gamma_\mu$ (summing over α and μ) and take the trace:

$$12 f_3 \left(\frac{p}{\Lambda}\right)^{2\epsilon} = \frac{-ie^2}{(2\pi)^4}(f_3 - f) \int \frac{d^4k}{k^4(p-k)^2} \left(\frac{k}{\Lambda}\right)^{2\epsilon} \left[12 p \cdot k(2 + 4g + 4g^2) \right.$$

$$- \frac{\psi}{k^2(p-k)^2} (\{k^4 + k^2 p^2 - 4gk^2 k \cdot (p-k) + 4g^2[k \cdot (p-k)]^2\} 12 p \cdot k$$

$$\left. + [-k^2 + 2gk \cdot (p-k)](16 p \cdot k^2 + 8p^2 k^2)) \right]. \quad (49)$$

This can be rearranged into the following useful form:

$$\left(\frac{p}{\Lambda}\right)^{2\epsilon} f_3 = \frac{-ie^2(f_3 - f)}{12(2\pi)^4} [12(2 + 4g + 4g^2)p_\mu J_2^\mu - \psi R_1 - 2g\psi R_2 - 48g^2 \psi R_3], \quad (50)$$

where $J_{2\mu}$ was given above and

$$R_1 = \int \frac{d^4k(k/\Lambda)^{2\epsilon}}{(p-k)^4 k^6} [2k^2(k^2 - p^2) + 2k^2(p^2 + k^2)(p-k)^2 - 4k^2(p-k)^4] = 0, \quad (51)$$

$$R_2 = \int \frac{d^4k(k/\Lambda)^{2\epsilon}}{k^6(p-k)^4} [4(p^2 - k^2)(p^2 + 2k^2) - 4(2p^2 - k^2)(p-k)^2 + 4(p-k)^4] k \cdot (p-k) = \frac{12i\pi^2 g}{(g-2)(g^2-1)} \left(\frac{p}{\Lambda}\right)^{2\epsilon}, \quad (52)$$

$$R_3 = \int \frac{d^4k p \cdot k[k \cdot (p-k)]^2}{k^8(p-k)^4} \left(\frac{k}{\Lambda}\right)^{2\epsilon} = \frac{i\pi^2}{2(g-1)(g-2)} \left(\frac{p}{\Lambda}\right)^{2\epsilon}, \quad (53)$$

where R_1, R_2, R_3 are evaluated in the manner used in the Appendix. Consequently,

$$f_3 = \frac{-\alpha}{4\pi} \frac{(f_3 - f)}{(g^2 - 1)} \left(2 + 4g + 4g^2 + \psi \frac{g^3}{g-2}\right) \quad (54)$$

or

$$\frac{f_3}{f} = \frac{\alpha[2 + 4g + 4g^2 + \psi g^3/(g-2)]}{4\pi(g^2 - 1) + \alpha[2 + 4g + 4g^2 + \psi g^3/(g-2)]}. \quad (55)$$

Having completed the calculation of f, g, ψ, and f_3, we substitute into Eq. (15) and evaluate F_1 as a function of α (or alternately g). The results will be discussed in Sec. IV.

IV. CONCLUSION

In this section we will show that our expression for F_1 reproduces low-order perturbation-theory results, discuss our solution in the light of gauge invariance, and discuss the singularity structure of F_1 as a function of α.

Our solutions will now be shown to agree with the results of low-order perturbation theory. We expand Eqs. (30) and (40) in a power series in the fine-structure constant and obtain

$$g = \frac{-\alpha}{4\pi} + \frac{1}{8}\left(\frac{\alpha}{2\pi}\right)^2 + O(\alpha^3), \quad (56)$$

$$\psi = \frac{-\alpha}{4\pi} + O(\alpha^2), \quad (57)$$

if we choose the branch of the equations corresponding to the results of low-order perturbation
theory. Substituting Eq. (55) into Eq. (15) yields

$$F_1 = \tfrac{2}{3}(1 - \tfrac{3}{2}g^2 - g^3)$$

$$- \frac{\alpha(2+4g+4g^2)(g-2)+\alpha\psi g^3}{4\pi(g^2-1)(g-2)+\alpha(2+4g+4g^2)(g-2)+\alpha\psi g^3}.$$

(58)

Since Eqs. (56) and (57) imply $g = \psi = 0$ at $\alpha = 0$,
we obtain

$$F_1 = \frac{2}{3} + \frac{\alpha}{2\pi} + (1+2g'-g'^2)\left(\frac{\alpha}{2\pi}\right)^2 + \cdots$$

(59)

when we expand around $\alpha = 0$, where

$$g' \equiv 2\pi \frac{dg}{d\alpha}\bigg|_{\alpha=0} = -\tfrac{1}{2}$$

by Eq. (56). Substituting for g' gives

$$F_1 = \frac{2}{3} + \frac{\alpha}{2\pi} - \frac{1}{4}\left(\frac{\alpha}{2\pi}\right)^2 + \cdots ,$$

(60)

in agreement with known exact results of low-
order perturbation theory. From Eq. (59) it is
clear that terms in Eq. (60) are only sensitive
to the $O(\alpha)$ behavior of g. This emphasizes the
importance of the functional dependence of Eqs.
(15) and (60) on g. This dependence results from
three assumptions: (1) power-law behavior of
Γ_μ, S_F, and $\Gamma_{\mu\alpha}$, (2) no contributions from terms
involving the electron-positron kernel, K, and
(3) the form of the vertex equation for $\Gamma_{\mu\alpha}$ given
in Eq. (9).

The dependence of g upon α can be understood
from the viewpoint of gauge invariance. From
Eq. (13) we see

$$Z_2 \sim \Lambda^{2\varepsilon}.$$

(61)

The generalized Landau gauge is defined[8] to be
the gauge where Z_2 is finite. The photon propagator in this gauge differs from the Feynman
gauge propagator by the term

$$\frac{Gk_\mu k_\nu}{k^4},$$

(62)

where $G = G(\alpha)$. Johnson and Zumino[9] have shown
that a change of gauge by

$$D_{\alpha\beta} \to D_{\alpha\beta} - \gamma \frac{k_\alpha k_\beta}{k^2}\left(\frac{1}{k^2+\mu^2} - \frac{1}{k^2+\Lambda^2}\right)$$

(63)

changes Z_2 by

$$Z_2 \to Z_2\left(\frac{\Lambda^2}{\mu^2}\right)^{(\alpha/4\pi)\gamma}.$$

(64)

In particular if we let $\gamma = -G + \psi$ and transform
from the generalized Landau gauge to the gauge

defined by ψ, we find

$$Z_2 \sim (\Lambda^2)^{(\alpha/4\pi)(\psi-G)}$$

(65)

in the ψ gauge. Using[8] $G = 1 - 3\alpha/8\pi$ and Eq. (57)
for ψ gives

$$\frac{\alpha}{4\pi}(\psi - G) = \frac{\alpha}{4\pi}\left(\frac{-\alpha}{4\pi} - 1 + \frac{3\alpha}{8\pi}\right)$$

(66)

$$= \frac{-\alpha}{4\pi} + \frac{1}{8}\left(\frac{\alpha}{2\pi}\right)^2.$$

(67)

Comparing Eqs. (56) and (67) establishes

$$g = \frac{\alpha}{4\pi}(\psi - G)$$

(68)

in low order. Therefore our equations embody
the general transformation properties of Z_2 under
gauge transformation, at least in low order.

Having established the agreement of our work
with low-order perturbation theory, we now
discuss the behavior of our approximation for
F_1. First, it should be noted that the solution for
g given above (which agrees with low-order perturbation theory) is only one of several possible
solutions. Figure 4 shows the relation between
α and g and displays the different branches of
$g(\alpha)$. The physical branch (i.e., the continuous
part corresponding to perturbation-theory results) of $g(\alpha)$ has the range $[0.61, -0.34]$ with
α varying from $-\infty$ to 3.4. The only singularity
of F_1 on the physical branch is a zero at $\alpha \approx -3.7$.
As Fig. 4 shows, the other branches have zeros
at $\alpha \approx -2$ and -3.6, and singularities at $\alpha \approx 0.9$
and -3.06. F_1 is a rapidly varying function of α
with a rich singularity structure in our approximation.

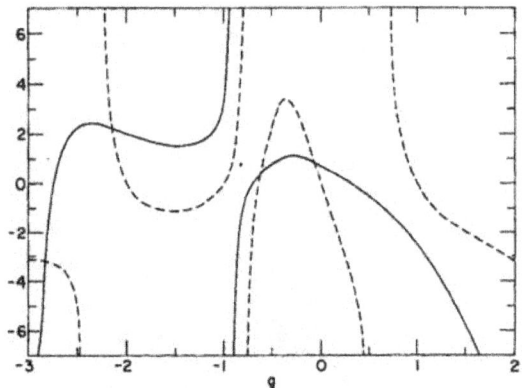

FIG. 4. The solid line is a parametric plot of F_1 versus
g. The dashed line is a plot of α as a function of g. To
find F_1 for a given α_0 draw a vertical line through the
point α_0 on the $\alpha(g)$ curve. The intercept on the F_1
curve gives $F_1(\alpha_0)$ for the branch chosen.

At this point it should be remarked that other choices for the argument of the zero-momentum-transfer vertex functions in the kernel of Eq. (6), $\Gamma_\lambda \cdots \Gamma_\sigma$, such as $\Gamma_\lambda(p) \cdots \Gamma_\sigma(p)$ will not lead to expressions for F_1 with essential singularities and can lead to forms for F_1 in disagreement with the low-order results of Eq. (60).

In conclusion, we have summed the contributions of $(2N+1)(2^N - N)$ diagrams to the $O(\alpha^N)$ terms of F_1 and (after summation over N) found that a rapidly varying function of α with branch-point singularities results.

ACKNOWLEDGMENT

It is a pleasure to thank Professor Marshall Baker and other members of the University of Washington Physics Department for helpful discussions. In addition, I am grateful to the Aspen Center for Physics for its hospitality and to the members of the Field Theory Workshop for interesting conversations.

APPENDIX

In this appendix we outline the method we have followed in evaluating loop integrals in Sec. III. Our procedure is to make a Wick rotation and express the loop integral in spherical coordinates in a four-dimensional Euclidean space. The following identities are useful[10]:

$$\frac{1}{1+\alpha^2 - 2\alpha t} = \sum_0^\infty C_N^1(t)\alpha^N \tag{A1}$$

for $\alpha < 1$ where C_N^1 is a Gegenbauer polynomial, and

$$\int d\Omega\, C_N^1 C_m^1 = 2\pi^2 \delta_{Nm}. \tag{A2}$$

All nontrivial angular integrals encountered in Sec. III reduce to one of the following forms:

$$\int \frac{d\Omega}{(p-k)^2} = \frac{2\pi^2}{k_>^2}, \tag{A3}$$

$$\int \frac{d\Omega}{(p-k)^4} = \frac{2\pi^2}{k_>^2(k_>^2 - k_<^2)}, \tag{A4}$$

where $k_>$ $(k_<)$ is the greater (lesser) of p and k.

We now will evaluate some integrals for the sake of illustration (we assume $g > 0$ throughout):

$$I_1 = \int \frac{d^4k}{(p-k)^2 k^2}\left(\frac{k}{\Lambda}\right)^{2\epsilon} \tag{A5}$$

$$= 2\pi^2 i \int_0^\Lambda \frac{dk\,k}{k_>^2}\left(\frac{k}{\Lambda}\right)^{2\epsilon} \tag{A6}$$

$$= 2\pi^2 i \left[\int_0^p \frac{dk}{p^2}k\left(\frac{k}{\Lambda}\right)^{2\epsilon} + \int_p^\Lambda \frac{dk}{k}\left(\frac{k}{\Lambda}\right)^{2\epsilon}\right] \tag{A7}$$

$$= \frac{i\pi^2}{g}\left[1 - \frac{1}{g+1}\left(\frac{p}{\Lambda}\right)^{2\epsilon}\right], \tag{A8}$$

which was given in Eq. (26). Our second example is

$$J_{0\mu\nu} = \int \frac{d^4k\, k_\mu k_\nu}{(p-k)^2 k^4}\left(\frac{k}{\Lambda}\right)^{2\epsilon}. \tag{A9}$$

Now

$$J_{0\mu\nu} = J_{01} g_{\mu\nu} + J_{02}\frac{p_\mu p_\nu}{p^2} \tag{A10}$$

and as a result

$$J_{0\mu}^\mu = 4J_{01} + J_{02} = I_1. \tag{A11}$$

Furthermore,

$$p_\mu p_\nu J_0^{\mu\nu} = p^2(J_{01} + J_{02}) \tag{A12}$$

$$= \int \frac{d^4k(k\cdot p)^2}{k^4(p-k)^2}\left(\frac{k}{\Lambda}\right)^{2\epsilon} \tag{A13}$$

$$= \frac{1}{4}\int \frac{d^4k}{k^4}\left(\frac{k}{\Lambda}\right)^{2\epsilon}\left[\frac{(p^2+k^2)^2}{(p-k)^2} - 2(p^2+k^2) + (p-k)^2\right] \tag{A14}$$

$$= \frac{\pi^2 i}{2}\left[\int_0^p \frac{dk}{k}\left(\frac{k}{\Lambda}\right)^{2\epsilon}\left(k^2 + \frac{k^4}{p^2}\right) + \int_p^\Lambda \frac{dk}{k}\left(\frac{k}{\Lambda}\right)^{2\epsilon}\left(p^2 + \frac{p^4}{k^2}\right)\right] \tag{A15}$$

$$= \frac{\pi^2 i}{4}\left[\frac{p^2}{g} + \frac{2}{g(g+2)(g^2-1)}\left(\frac{p}{\Lambda}\right)^{2\epsilon} - \frac{4}{(g+2)(g-1)}\right] \tag{A16}$$

up to terms of $O(p^2/\Lambda^2)$, which we drop. Algebraic manipulation of Eqs. (A16) and (A10) gives

$$J_{01} = \frac{i\pi^2}{4}\left[\frac{1}{g} + \frac{2}{g(g+2)(g^2-1)}\left(\frac{p}{\Lambda}\right)^{2\epsilon}\right]. \tag{A17}$$

$$J_{02} = \frac{-i\pi^2}{(g+2)(g-1)}\left(\frac{p}{\Lambda}\right)^{2\epsilon}. \tag{A18}$$

The above equation leads directly to Eq. (34).

*Work supported in part by the U.S. Atomic Energy
  Commission.
†Address after September 1, 1973: Physics Department,
  Cornell University, Ithaca, New York 14850.
[1]M. Baker and K. Johnson, Phys. Rev. D 3, 2516 (1971),
  and references contained therein; K. Johnson and
  M. Baker, ibid. 8, 1110 (1973).
[2]S. Adler, Phys. Rev. D 5, 3021 (1972).
[3]J. L. Rosner, Phys. Rev. Lett. 17, 1190 (1966); R. Jost
  and J. M. Luttinger, Helv. Phys. Acta 23, 201 (1950);
  A. E. Uehling, Phys. Rev. 48, 55 (1935).
[4]H. Brandt, thesis, University of Washington (Seattle)
  (unpublished). I am grateful to Professor M. Baker

for drawing my attention to this article.
[5]K. Johnson, R. Willey, and M. Baker, Phys. Rev. 163,
  1699 (1967).
[6]S. Blaha, Univ. of Washington report, 1972 (unpublished).
[7]M. Gell-Mann and F. Low, Phys. Rev. 95, 1300 (1954).
[8]K. Johnson, M. Baker, and R. Willey, Phys. Rev. 136,
  B1111 (1964).
[9]K. Johnson and B. Zumino, Phys. Rev. Lett. 3, 351
  (1959).
[10]W. Magnus and F. Oberhettinger, Formulas and Theo-
  rems for the Functions of Mathematical Physics
  (Chelsea, New York, 1949).

PHYSICAL REVIEW D   VOLUME 9, NUMBER 8   15 APRIL 1974

Higher-order calculation of transmission below the potential barrier

S. S. Wald and P. Lu

Department of Physics, Arizona State University, Tempe, Arizona 85281
(Received 26 December 1973)

In extending the Miller-Good modified WKB approximation to include the higher-order terms, a
divergence was introduced. Because of this divergence, the approximation was limited to energies above
the potential barrier. With this divergence removed, the modified WKB method is no longer limited to
energies above the potential barrier. In order to demonstrate this method, we calculate the transmission
coefficients for energies below the peak of the potential barrier and show that the higher-order terms
are essential to the approximation.

I. INTRODUCTION

The conventional WKB approximation is widely
known for its usefulness in solving simple barrier-
penetration problems. However, as Ford et al.[1]
pointed out, the conventional WKB method tends to
break down as the energy approaches the potential-
barrier top.

Miller and Good[2] proposed a modified WKB method in which the solutions of a model Schrödinger
equation that can be solved exactly and resembles
the actual Schrödinger equation would be used as
the basis of the approximation. The reader is referred to their paper for details. However, the
modified WKB method was only utilized to zeroth
order in $\hbar^2$ because of divergences in the higher-
order terms. Using the method developed by Lu
and Measure[3] to remove the divergences in the
higher-order terms, a divergence at the maximum
point of the potential barrier was introduced. This
divergence limits the approximation to energies
above the barrier top where there are no real
classical turning points, and hence there is no
maximum point on the path of integration. Using
the modified WKB method to first order in $\hbar^2$, we
calculated the transmission coefficients above the
potential barrier[4] and obtained agreement with the

numerical results to at least four significant figures. This indicated how essential the higher-order terms are to the approximation.

For the case of penetration below the potential
barrier, there are two classical turning points
and one maximum point lying between the turning
points. In order to remove the divergence at the
maximum point, we start with the basic contour
integral representation and then derive a formula
which can be applied to the case of "penetration
through the barrier." In Sec. II, this formula is
derived in general terms. Using the Eckart potential as an example in Sec. III, we calculate the
transmission coefficients for energies below the
barrier, and the results are shown to be in agreement with the numerical results as presented in
Table I. Thus the barrier-penetration problem
can be solved using the modified WKB method with
excellent results even for energies near the top
of the potential barrier.

II. METHOD OF APPROXIMATION

In general, we wish to solve the Schrödinger
equation

$$\left[\frac{d^2}{dx^2} + \frac{P_1{}^2(x)}{\hbar^2} \right] \psi(x) = 0 \tag{1}$$

3. Temporal View of QED Vacuum Polarization Distribution

From eqs. 65 – 68 of Blaha's calculation of the Johnson-Baker-Willey eigenvalue function we see that the renormalization constant Z_3, which relates the bare electron charge α_0 to the renormalized charge α:

$$\alpha = Z_3\alpha_0 \tag{3.1}$$

has the form

$$Z_3 \sim \Lambda^{2g} \tag{3.2}$$

where Λ is a large mass and g is a constant determined by α in Appendix 2-A and listed in Table 2.1: The momentum dependent form of Z_3 is thus

$$Z_3 = \Gamma(p) = (p^2/\Lambda^2)^g \tag{3.3}$$

up to a constant by eq. 2-A.9. The physical picture of the vacuum polarization consists of an (infinite) bare charge α_0 that is masked by vacuum sea electrons that are repulsed from the bare charge electromagnetically. The displacement of the sea electrons is equivalent to the presence of a positive charge distribution. The result is that at long distances the positive charge distribution almost compensates for the infinite negative bare charge resulting in the fine structure constant that we measure $\alpha = 1/137\ldots..$

If we probe an electron at high energy the apparent charge of the electron is more negative. At very large momentum the electron charge becomes enormously negative. $\Gamma(p)$ displays this behavior for a negative value of g, which we found to be the case in Blaha (2019f) and Table. 2.1:

$$g = -\,0.00058053691948 \tag{3.4}$$

The momentum p is a measure of the energy E, at which Z_3 is being probed. Since we are considering vacuum fluctuations that generate Z_3 the energy-time uncertainty condition is relevant:

$$\Delta E\,\Delta t \geq \tfrac{1}{2}\hbar \tag{3.5}$$

It implies pairs of particles with energy ΔE have a lifetime less than Δt. Thus a vacuum fluctuation polarization of energy ΔE has a lifetime approximately equal to Δt. Large p behavior is associated with short lifetime vacuum fluctuations. We can see the time behavior by Fourier transforming $\Gamma(p)$:

$$\Gamma(t) = \int\limits_{0}^{\infty} d(\ln p)\ e^{-ipt}\ \Gamma(p)/(2\pi)^{\frac{1}{2}} \tag{3.6}$$

$$= k\ (t/T)^{-2g} \tag{3.7}$$

where k is a constant[14]

$$k = \Gamma(2g)e^{-i\pi g}/(2\pi)^{\frac{1}{2}} \tag{3.8}$$

and where T is a large time with

$$\Lambda = 1/T \tag{3.9}$$

The small t behavior of the vacuum polarization corresponds to the large (high energy) p behavior. The large t behavior of the vacuum polarization corresponds to the small (infrared) p behavior.

In chapter 4 we will consider a universe particle produced by the annihilation of a fermion-antifermion pair. The universe particle will acquire a "vacuum polarization" from the Dirac sea of universe particles. We will calculate the vacuum polarization of the universe particle and obtain vacuum polarization expressions similar to those seen above.

We then will turn to examining the scale factor a(t) in relation to the Fourier transform of the vacuum polarization $\Gamma(t)$. We will then see a remarkable correspondence.

---

[14] $\Gamma(2g)$ is the mathematics gamma function (generalized factorial function). Its argument 2g is a negative number.

4. Vacuum Polarization and Universe Expansion

4.1 Universe Vacuum Polarization - A New Vector Interaction for Universe Particles

We assume universes can be treated as particles in 4-dimensional space-time.[15] Since experiments appear to have shown that the universe does not rotate (does not have spin)[16] we will assume the universe is a spin 0 boson. We also assume that universes have a vector field interaction similar to QED. It is possible that the quantum vector $Y^\mu(x)$ field of the Big Bang quantum coordinates[17] may be the vector field universe interaction field[18].

Given this QED-like framework, universe-antiuniverse pair production and vacuum polarization becomes possible. We assume the QED-like lagrangian

$$\mathcal{L} = \tfrac{1}{2}\,(\partial_\mu\varphi^\dagger\partial^\mu\varphi - m^2\varphi^\dagger\varphi) - ie_0\colon \varphi^\dagger(\overrightarrow{\partial_\mu} - \overleftarrow{\partial_\mu})\,\varphi\colon A^\mu + e_0^2\colon\!A^2\!\colon \colon\!\varphi^\dagger\varphi\colon + \delta m^2\colon\!\varphi^\dagger\varphi\colon$$

(4.1)

where $\varphi(x)$ is a "charged" quantum universe particle field and A^μ is a QED-like vector field.[19]

Our scenario starts with the creation of a universe particle through the annihilation of a fermion-antifermion pair in the Megaverse. The universe particle produced is a scalar. It expands as a fireball—not as jets of particles—but as a subspace with matter (particles) bound together by gravitation due to an enormous mass-energy.

The expanding universe has a vacuum polarization due to an electromagnetic-like field A^μ that yields a similar to QED-like vacuum polarization. The universe vacuum polarization mirrors the expansion of the universe.

We now proceed to calculate the second order vacuum polarization of a universe particle. Then we will use it to determine the universe eigenvalue function, and the resulting universe fine structure constant and the universe power g_U.

4.2 Second Order Vacuum Polarization of a Universe Particle

The relevant one loop vacuum polarization Feynman diagram appears in Fig. 4.1. Its evaluation is:

$$I_{\mu\nu} = (-ie_0)^2 \int \frac{d^4k}{(2\pi)^4}\,\frac{i}{(k^2 - m^2 + i\varepsilon)}\,\frac{i}{(k^2 - m^2 + i\varepsilon)}(q - 2k)_\mu(q - 2k)_\nu \qquad (4.2)$$

---

[15] Universes are composite entities but we can treat them as quantum particles in the same manner as physicists treated protons and neutrons etc. as quantum particles before quark theory was accepted.

[16] The lack of universe rotation (spin) is indicated by a study of Cosmic Microwave Background (CMB) by D. Saadeh *et al*, Phys. Rev. Lett. **117**, 313302 (2016).

[17] Blaha (2019e).

[18] Or the field may be related to the *vierbein* field of General Relativity.

[19] The charge is not electromagnetic charge.

$$= \frac{\alpha}{2\pi} \int_0^\infty dz_1 \int_0^\infty dz_2 \frac{g_{\mu\nu} \exp[i(q^2 z_1 z_2/(z_1 + z_2) - (m^2 + i\varepsilon)(z_1 + z_2))]}{(z_1 + z_2)^3} + \text{gauge terms}$$

upon introducing parameters z_1 and z_2 to enable exponentiation and integration over k, where

$$\alpha = e_0^2/4\pi \tag{4.3}$$

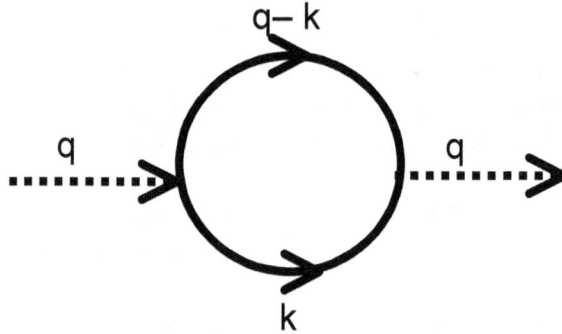

Figure 4.1 One loop vacuum polarization boson Feynman diagram.

Applying $q^2 \partial/\partial q^2$ to $I_{\mu\nu}$ to eliminate the quadratic divergent part, and then using the identity

$$1 = \int_0^\infty d\lambda/\lambda \ \delta(1 - (z_1 + z_2)/\lambda)$$

and letting $z_i = \lambda x_i$ we obtain

$$I_{\mu\nu} = \frac{i\alpha}{2\pi} q^2 g_{\mu\nu} \int dx_1 \int dx_2 \int d\lambda/\lambda \ x_1 x_2 \exp[i\lambda(q^2 x_1 x_2 - (m^2 + i\varepsilon))] \ \delta(1 - x_1 - x_2)$$

$$\tag{4.4}$$

up to gauge terms. The λ integration yields a logarithmic divergence which we cut off. Then

$$I_{\mu\nu} = \frac{i\alpha}{2\pi} q^2 g_{\mu\nu} \int_0^1 dx \ x(1-x) \ln(q^2 x(1 - x) - m^2) \tag{4.5}$$

which becomes

$$I_{\mu\nu} = \frac{i\alpha}{12\pi} q^2 g_{\mu\nu} \ln(\Lambda^2/m^2) + \ldots \tag{4.6}$$

with finite terms not shown.

Thus we find the renormalization constant Z_3 for the scalar universe particle case to be

$$Z_3 = 1 - \alpha/12\pi \ln(\Lambda^2/m^2) \tag{4.7}$$

If we let

$$\alpha_U = \alpha/4 \tag{4.8}$$

then we obtain the form similar to the one loop value of Z_3 for spin ½ electron QED:[20]

$$Z_3 = 1 - \alpha_U/3\pi \ln(\Lambda^2/m^2) \tag{4.9}$$

We now provisionally assume that α is the QED fine structure constant. We denote it as α_{QED}. Note $\alpha_{QED} = 0.0072973525693$ (11)

Thus the "fine structure constant" α_U for our vector interaction is

$$\alpha_U \equiv \alpha_{QED}/4 = 0.001824338 \tag{4.10}$$

We now turn to the Johnson-Baker-Willey (JBW) model of massless QED since at ultrahigh energy our vector interaction theory with lagrangian eq. 4.1 becomes the JBW model. In the JBW model we calculated α_{QED} and found the corresponding power which we denote g_{QED}. Next we perform the same calculation and find the g value denoted g_U corresponding to α_U. The value of g_U will be seen to lead to the power g in the universal scale factor almost exactly.

4.3 Application of the Approximate Coupling Constant Calculation to Universe Vacuum Polarization

We now extend the JBW eigenfunction determination to the case of the vacuum polarization of a universe particle described by

$$(\alpha_U/2\pi) = [g_U A_4(g_U) - (4 + 2g_U)A_2(g_U)]/(A_4(g_U)A_1(g_U) - A_2(g_U)A_3(g_U)) \tag{4.11}$$

since the second order form of Z_3 in eq. 4.9, which is the same as the QED second order form of Z_3, generalizes to all orders as a function of α_U.

Given the value of

$$\alpha_U = 0.001824338 \tag{4.12}$$

in eq. 4.10 we can extract the value of g_U by inverting eq. 4.11 to obtain g_U:

$$g_U = -0.00014525 \tag{4.13}$$

4.4 Relating Vacuum Polarization to the Universal Scale Factor

We now relate the vacuum polarization found above to the growth of the universe as given by the universe scale factor:

---

[20] The factor of 4 originates in the spin 0 nature of a universe particle in contrast to the spin ½ nature of QED electrons.

$$\Gamma_{\mu\alpha}(p) = (f_3/p^2)(\gamma{\cdot}p\gamma_\mu\gamma_\alpha - \gamma_\alpha\gamma_\mu\gamma{\cdot}p)(p/\Lambda)^{-2g_U}$$

where we define the vacuum polarization factor

$$\Gamma_U(p) = (p/\Lambda)^{2g_U} \tag{4.14}$$

up to a constant factor. We now fourier transform $\Gamma_U(p)$ to the time representation t as in eq. 3.6 for QED vacuum polarization:[21]

$$\Gamma_U(t) = \int_0^\infty d(\ln p)\ e^{-ipt}\,\Gamma_U(p)/(2\pi)^{\frac{1}{2}} \tag{4.15}$$

$$= k\ (t/T)^{-2g_U} \tag{4.16}$$

where k is a constant[22]

$$k = \Gamma(2g_U)e^{-i\pi g_U} \tag{4.17}$$

and where T is a large time with

$$\Lambda = 1/T \tag{4.18}$$

The small t behavior of the vacuum polarization corresponds to the large (high energy) p behavior. The large t behavior of the vacuum polarization corresponds to the small (infrared) p behavior.

---

[21] Note the correspondence to eqs. 3.6 and 3.7.

[22] $\Gamma(-2g)$ is the mathematics gamma (generalized factorial) function. Its argument 2g is a negative number.

5. Phenomenological Scale Factor a(t)

In chapter 4 we saw that

$$\Gamma_U(t) = k\,(t/T)^{-2g_U} \tag{4.16}$$

where k is a constant. In eq. 10.9 of Blaha (2021b) we defined a phenomenological universe scale factor as

$$a(t) = [(t + t_0)/t_{now}]^{g\,[1 + d/(t + t_0)] + h(t + t_0)} \tag{5.1}$$

$$= [(t + t_0)/t_{now}]^{gd/(t + t_0)} [(t + t_0)/t_{now}]^{g}\, [(t + t_0)/t_{now}]^{h(t + t_0)} \tag{5.2}$$
$$\textbf{I} \qquad\qquad\qquad \textbf{II} \qquad\qquad \textbf{III}$$

Where the labels I, II, and II indicate the role of each factor in the time evolution of the universe: I indicates the Big Bang period, II indicates an intermediate period, and III indicates the "recent" period.

In this chapter we will see that the intermediate period factor II corresponds qualitatively, and numerically, to the Fourier time transform of the universe vacuum polarization calculated in chapter 4.

Appendix 5-A describes the motivation, and the numerical determination, of the phenomenological fit. The author believes that the attempts to create a Standard Model of Cosmology will eventually to a scale factor that is similar to the phenomenological fit. The fit has some remarkable features such as a Big Dip that appears to correspond to the transformation of the evolving universe to a matter dominated state.

5.1 Particle Universe Scenario

In Blaha (2021b) we pointed to theoretical and experimental data that supported the view that the universe, indeed any universe, is a type of particle that differs from elementary particles in that it has a subspace and that it dynamically expands.

Based, in part, on the Octonion Cosmology theory we believe universes are created by fermion-antifermion annihilation in a parent space. The created universe has its own space and dimensions as well as mass and energy. The creation event may be simulated in quantum field theory as we did in Blaha (2021a) and (2021c). It seems to necessarily be an event marked by a singularity. In our phenomenological fit to a(t) we simulate the creation event with an essential singularity that has been suggested to mark the beginning of the Big Bang.

Subsequent to the creation event and a Big Bang period (corresponding to period I in eq. 5.2), the universe proceeds to expand along the lines of the universe vacuum polarization (corresponding to period II in eq. 5.2) calculated in chapter 4. Then the universe enters a new phase (corresponding to period III in eq. 5.2) with a Big Dip and further expansion. Thus one has a global view of the expansion of the universe from the beginning.

5.2 Universe Vacuum Polarization and the Phenomenological Universe Scale Factor

In chapter 4 we determined the power in $\Gamma_U(t)$ to be

$$g_U = -0.00014525 \tag{4.13}$$

where

$$\Gamma_U(t) = k \, (t/T)^{-2g_U} \tag{4.16}$$

In appendix 5-A we determined the power in region II to be

$$g = 0.000282377 = 2.82377 \times 10^{-4} \tag{5.3}$$

where

$$a(t) \cong (t/t_{now})^{g + ht} \cong (t/t_{now})^g \tag{5.4}$$

Identifying T with t_{now} , since they both correspond to the maximum time, and noting t_0 is negligible in regions II and II we see that we may equate

$$g = -2g_U$$
$$= 0.0002905 \tag{5.5}$$

to an accuracy of less than 3% with agreement as well in the signs of g and $-2g_U$. This remarkable agreement supports the vacuum polarization interpretation of universe expansion, and the interpretation of universes as particles.

Given the approximate nature of our calculations of vacuum polarization the agreement is remarkable.[23]

The dependence of the universal scale factor on g governs the "small" region II time behavior of the universe. Correspondingly, the dependence of the vacuum polarization on g_U is a large momentum phenomena.

The parameter h in a(t) is set primarily by the larger t (region III times) behavior of a(t). It corresponds to the infrared behavior of its fourier transform $\Gamma_U(p)$ when the infrared (possibly mass dependent) behavior of the vacuum polarization is calculated.

The preceding discussion also demonstrates that our earlier assumption

$$\alpha_U \equiv \alpha_{QED}/4 = 0.001824338 \tag{4.10}$$

---

[23] And may be exact! The value of the Hubble Constant H in recent times varies from about 70 – 75 making the calculation of g also approximate. We chose an average value of 73.24 to obtain the value of g above. If we chose the current value for H to be 75.58 we would have $g = -2g_U$ exactly. Note: studies of binary black hole merger gravity waves have given a Hubble Constant of 75.2 km s^{-1} Mpc^{-1} (and earlier of 78 km s^{-1} Mpc^{-1}), and studies of light bent by distant galaxies give H = 72.5 km s^{-1} Mpc^{-1}. Thus the value H = 75.58 is not unreasonable.

is correct, since it leads directly to the power g in the universal scale factor. Thus the evolution of our universe is partly set by the vacuum polarization.

We thus have the growth rate g_U and the coupling constant α_U for the universe.

5.3 Vacuum Polarization Interpretation of the Universal Scale Factor

The vacuum polarization view of the time evolution of the universe requires that we view the entire time evolution of the universe as a whole. Normally one views time as increasing. Feynman has suggested we could also view time as flowing backward.

We now have a new view where we consider the life history of the universe as a static event rather like a time lapse picture of a flower's growth. The thought process is similar to that of Feynman path integral formulations, which consider the complete path in time of a process.

5.4 Possible Exact Match of Universe Vacuum Polarization and Universe Expansion

We found the vacuum polarization for a *scalar* universe above. The growth rate of the vacuum polarization g_U in region II was ½ of the growth rate of universe expansion g. There are two cases, in which $g_U = g$ results: the case of a spin ½ universe (although no spin has been detected in our universe), and the case where the scalar universe has charge 2e (twice the electromagnetic charge value) although the universe particle charge is not electric charge.

With either of these cases the time dependence of the vacuum polarization equals the time dependence of the scale factor a(t) in region II.

5.5 The Dirac Particle Sea vs. the Universe Mass-Energy Expansion

The possibilities raised above in sections 5.3 and 5.4 yield a comparative picture of vacuum polarization and universe expansion in region II. The expansion of the universe, which takes place above the Dirac sea, equals the displacement of the vacuum polarization within the Dirac sea.[24]

A justification of the correspondence is apparent.

5.6 Regions I and III of a(t)

In eq. 5.1 and 5.2, and section 5-A.4 of the appendix 5-A, we describe the new expanded form of the scale factor a(t).

$$a(t) = [(t + t_0)/t_{now}]^{gd/(t + t_0)}[(t + t_0)/t_{now}]^{g} [(t + t_0)/t_{now}]^{h(t + t_0)} \qquad (5.2)$$
$$\text{I} \qquad\qquad\qquad \text{II} \qquad\qquad \text{III}$$

We found region II matched the universe vacuum polarization. The other regions may be viewed as having analogs in the universe vacuum polarization *IF* an exact calculation were performed.

---

[24] The displacement can be viewed as virtual particles of the same sign in the Dirac sea being "pushed away" from the charge. Or it can be viwed as virtual particles of the opposite sign in the Dirac sea being "attracted" to the central charge. Thus the "build up" of the charges below in the Dirac sea equals the "build up" of mass-energy above the Dirac sea in universe expansion.

5.6.1 Region I Vacuum Polarization Equivalent

The region I a(t) is approximately

$$a_I(t) = [(t + t_0)/t_{now}]^{gd/(t + t_0)} \tag{5.6}$$

This region may be viewed as a subregion where $0 \leq t \leq t_0$, and as another subregion where $-t_0 < t < 0$. In the first subregion, which is part of the Big Bang region, we see

$$a_I(t) \cong [t_0/t_{now}]^{gd/t_0} \tag{5.7}$$

is constant to good approximation. Note a(0) is a finite, non-zero number.

In the second subregion, which is also part of the Big Bang region, a(t) varies up to an infinite value at $t = -t_0$. We view the divergence, an essential singularity, as an indication of a transition from a fermion-antifermion pair to a universe particle in Octonion Cosmology. Its analog is an essential singularity in the ultraviolet region in the vacuum polarization that was conjectured by Adler. (See section 5-A.4) The vacuum polarization singularity appears to be beyond perturbative calculation. See chapter 6 for additional detail.

5.6.2 Region III Vacuum Polarization Equivalent

The region III a(t) is approximately

$$a_{III}(t) = [(t + t_0)/t_{now}]^{g + h(t + t_0)} \tag{5.8}$$
$$\cong [t/t_{now}]^{g + ht} \tag{5.9}$$

It is governed primarily by the "ht" exponent. The implicit logarithms in the $t^{ht} = e^{ht \ln t}$ factor suggest that the corresponding vacuum polarization has an infrared part related to soft photon-like behavior seen in QED. This feature of vacuum polarization is not present in the "high energy" vacuum polarization calculation described here.

5.7 Higher Spin Universes and Megaverses

The discussions in this, and the previous, chapters can be generalized to cases where there are higher spin universes and Megaverses.

We conclude with the view that the phenomenological scale factor a(t) is likely to be an accurate depiction of a(t) calculated in The Standard Model of Cosmology and in other possible scenarios.

Appendix 5-A. The Phenomenological a(t)

This appendix describes the motivation for the phenomenological scale factor a(t) and its numeric calculation. The basis of the phenomenological fit is:

1. Power law behavior (in part) as in the radiation and matter dominated approximations for a(t).

2. The known general shape of H(t) at early times, and at present: a massive decline from the Big Bang period and a recent rise.

3. The simplicity of the model. Two values of H(t) set the constants g and h.

4. Faster than exponential future growth as $t \to \infty$.

$$a(t) = \exp[(g + ht)\ln(t/t_{now})] \sim e^{ht\,\ln(t)}$$

5. *The small time behavior of a(t) can be derived in a particle model of a universe as shown in detail later in Blaha (2019c).*

6. The determination of the universe vacuum polarization phase of universe growth.

7. The Adler[25] conjecture of an essential singularity in the universe vacuum polarization, which can be used to determine the Big Bang beginning.

Universes and Megaverses grow from annihilation events in a Megaverse and a Maxiverse respectively. In this appendix we will consider the growth of universes and Megaverses (and possibly other space instances) from the initial annihilation based on the assumption of a generalization of coordinates in the created instance to Two Tier quantum coordinates that the author developed many years ago. Quantum coordinates "smear" coordinates to prevent divergences in quantum field theory calculations,[26] and, in the present case, to prevent a catastrophe at the "Big Bang" point of the instance.[27]

Later we will consider a generalization of the formula for the universe's scale factor and Hubble Parameter to accommodate the Big Bang period as well as more recent times. This formula can be based on a universe/Megaverse model with a "vacuum polarization" that is analogous to that in Blaha's exact calculation of the Fine

---

[25] S. Adler, Phys. Rev. **D5**, 3021 (1972).
[26] Blaha (2002), and (2005a).
[27] Blaha (2004) and (2019e).

Structure Constant, α, based on vacuum polarization in the Johnson-Baker-Willey formulation of Quantum Electrodynamics. *It is a model based on a close analogy (conceptually and numerically) with the Quantum Electrodynamics vacuum polarization of a particle. We believe an octonion space instance is a type of particle.*

Our interim Big Bang Model was based on a quantum coordinate that is defined using a massless vector field denoted $Y^\mu(y)$:

$$X^\mu(y) = y^\mu + i\ Y^\mu(y)/M_c^2 \qquad (5\text{-A.1})$$

where the y^μ coordinates are ordinary "point" coordinates.

In Blaha (2019e) we calculated the expansion of a universe (or a Megaverse) from the "Big Bang" point and showed the "quantum pressure" due to the Planck Black Body energy distribution of the $Y^\mu(y)$ fields prevented an infinite collapse at t = 0. We then matched the scale factor and Hubble Parameter of our Big Bang model with the empirically found values that we found it described the universe subsequently. Here we extend our empirical model to the Big Bang of this, and other octonion space, instances.

5-A.1 Growth Scenario

Growth begins with the transition of the fermion-antifermion pair to a universe (or Megaverse) particle. The fermion-antifermion pair use Two-Tier quantum coordinates and the created particle's coordinates are TwoTier coordinates in its interior.

The particle contains an extremely large internal energy distribution that primarily resides in the Planck Black Body energy distribution of the $Y^\mu(y)$ fields. The energy distribution initially stabilizes the particle to a finite size preventing a collapse at t = 0.

The particle then expands in a Big Bang generating the growth of the particle (universe or Megaverse).

5-A.2 The Blaha (2019a) Calculation of Growth

The calculation of the growth of the universe in the interim model[28] showed an enormous growth as expected.

The radius of the universe is non-zero as expected due to the use of Two-Tier Quantum coordinates. At t =0

$$r = 4.278 \times 10^{-65} \text{ cm} \qquad (5\text{-A.2})$$

and at time $t_c = 1.26 \times 10^{-165}$ s

$$r = 8.5 \times 10^{-65} \text{ cm} \qquad (5\text{-A.3})$$

with an effective doubling of the radius.

---

[28] Sections 5-A.2- 5-A.4 quotes Blaha (2019c)'s text and equations using its equation numbers.

The scale factor a(t) at time t = 0 is

$$a(0) \cong 3.19 \times 10^{-93}$$ (5-A.4a)

and at $t_c = 1.26 \times 10^{-165}$ s is

$$a(t_c) \cong 1.632 \times 10^{-92}$$ (5-A.4b)

The Hubble Parameter at t = 0 and $\check{r}$ = 1, the scaled radius at the edge of the Big Bang, is

$$H_{BBRW}(0, 1) \cong 1.79 \times 10^{218} \text{ km s}^{-1} \text{ Mpc}^{-1}$$ (13.4.7)

(compared to the current Hubble constant of 100h km s^{-1} Mpc^{-1}.
At t = 0 and $\check{r}$ = 0,

$$H_{BBRW}(0, 0) \cong 2H_{BBRW}(0, 1)$$ (13.4.8)

At the "edge" of the universe at time $t_c = 1.26 \times 10^{-165}$ s and $\check{r}$ = 1 the Hubble Parameter is[29]

$$H_{BBRW}(t_c, 1) \cong 1.14 \times 10^{126} \text{ km s}^{-1} \text{ Mpc}^{-1}$$ (13.4.9)

At at time $t_c = 1.26 \times 10^{-165}$ s and $\check{r}$ = 0, the "center" of the Big Bang has

$$H_{BBRW}(t_c, 0) \cong 8.95 \times 10^{217} \text{ km s}^{-1} \text{ Mpc}^{-1}$$ (13.4.10)

using eq. 13.3.2.14. *The "center" is more rapidly expanding by a factor of the order of 10^{93}, which is suggestive of a developing, much less dense region.*

5-A.2.1 Creation of a Void in the Metastable State due to a Radially Varying Hubble Constant

Sections 13.6.1 and 13.6.2 show that at t = 0 the Hubble Parameter is a factor of 2 greater at $\check{r}$ = 0 than at the edge of the universe ($\check{r}$ = 1), and at t = t_c (the end of the metastable state life) the Hubble Parameter is a factor of the order of 10^{93} greater at $\check{r}$ = 0 than at the edge of the universe ($\check{r}$ = 1). The vast disparity in Hubble Parameter values at the center compared to the edge of the universe suggests the center will rapidly go to lower density creating a *void*. *Thus one can expect a "Hubble bubble" generated during the metastable state.*

Today we see numerous voids in the universe.[30] The largest is the spherical KBC supervoid (containing the Milky Way) with a 2 billion light year diameter. Another supervoid is the "Giant void" (Canes Venatici) with a diameter of 1.3 light years. There is also the WMAP "cold spot" void with a diameter of 120 Megaparsecs.

---

[29] Using eq. 13.3.2.15. of Blaha (2019e)
[30] Zehavi *et al*, Astrophysical Journal **603**, 483 (1998), There are differing results on the Hubble Bubble question such as in Moss *et al*, Phys. Rev. **D83**, 103515 (2011) and references therein.

The Hubble Parameter within this void is larger due to the attraction of mass-energy external to the void just as we see above when comparing the center to the edge of the Metastable state.

The metastate "void" appears to be a precursor to the later voids seen now. Voids may also be the result of the fluctuation in universe size around the time of the Big Dip.

5-A.3 Original Phenomenological Model to the Hubble Parameter

Another calculation of a model for the change of the Hubble Parameter over time used the experimentally determined values

$$H(t_c) \equiv H(380{,}000 \text{ yr}) = 67.8 \qquad\qquad (5\text{-A.5})$$
$$H(t_{now}) = 73.24$$

These values approximate the set of known experimentally determined values. The phenomenological model was based on the assumptions:

The Hubble Constant implied by the phenomenological a(t) of eq. 5.1is

$$H(t) = (da/dt)/a = g/t + h(1 + \ln(t/t_{now})) \qquad\qquad (22.2)$$

If we set the value of H(t) at two values of time, then g and h are determined. Based on the above discussion we use the experimental data:

$$H(380{,}000 \text{ yr}) = 67.8 \qquad\qquad (22.3)$$
$$H(t_{now}) = 73.24 \qquad\qquad (22.4)$$

Eqs. 22.2 and 22.3 imply

$$h = (t_c H(t_c) - t_{now}H(t_{now}))[\, t_c - t_{now} + t_c \ln((t_c/t_{now})]^{-1} \qquad (22.5)$$
$$g = (H(t_{now}) - h)\, t_{now}$$

Substituting the parameter values of eq. 22.3 we obtain[31]

$$h = 2.25983 \times 10^{-18} \qquad\qquad (22.6)$$

$$g = 0.000282377 = 2.82377 \times 10^{-4}$$

Since

$$h \cong 1/t_{now}$$

an alternate possible form for a(t) is

$$a(t) = (t/t_{now})^{g + t/t_{now}} \qquad\qquad (22.1a)$$

---

[31] The calculation of g, in particular, is delicate since it contains small differences between large quantities.

We found this form to be suggestive but not consistent with the current values of the Hubble Parameter.

If one wanted the value of h to be equal to $1//t_{now}$ then the present value of the Hubble Parameter would have to be 74.47 (if the Hubble value at 380,000 years is 67.8) – a value within the range of experimental values."

5-A.3.1 Original Model a(t) at Small Times (the Big Bang Period)

The original model for a(t), eq. 22.1a, seems to work well for large times greater than 380,000 years. However at very small times such as the Big Bang period a(t) must be modified since

$$a(0) = 0 \qquad \text{BAD} \qquad (5\text{-A.6})$$

at t = 0 resulting in a divergence. H(t) is also divergent at t = 0.

5-A.4 A New Phenomenological Model Extending to Big Bang Period

If we examine a(t) for small times we find eq. 22.1a above we find

$$a(t) \sim (t/t_{now})^g \qquad (5\text{-A.7})$$

where the constant g can be determined by a vacuum polarization calculation as seen in Blaha (2019c). Furthermore the numeric value of g is also determined in that calculation.

A study of the Johnson-Baker-Willey model by Adler[32] suggested that the vacuum polarization summed to all orders might have an essential singularity at α perhaps of the form

$$\exp[-1/(\alpha - 0.0072973525693)] \qquad (5\text{-A.8})$$

where α has the known 0.0072973525693.[33]

Motivated by this suggestion, and the successful understanding of eq. 5-A.7 from vacuum polarization, we propose a(t) has an additional factor that governs its behavior in the Big Bang period:[34]

$$a(t) \cong [(t + t_0)/t_{now}]^{g + h(t + t_0)} [(t + t_0)/t_{now}]^{gd/(t + t_0)} \qquad (5\text{-A.9})$$
$$= [(t + t_0)/t_{now}]^{g\,[1 + d/(t + t_0)] + h(t + t_0)}$$

where t_0 is a base time value, and d is a constant. The new a(t) has an essential singularity at the unphysical $t = -t_0$. Since t_0 will be seen later to be extremely small it is possible to view the state at $t = -t_0$ as a precursor state of a fermion-antifermion annihilation as we did earlier in Octonion Cosmology.

The resulting Hubble Parameter is

---

[32] S. Adler, Phys. Rev. **D5**, 3021 (1972).
[33] Calculated to the exact known experimental value in Blaha (2019f).
[34] We choose $t + t_0$ rather than $t - t_0$ to avoid an essential singularity at positive t.

$$H(t) = [g + h(t + t_0) + gd/(t + t_0)]/(t + t_0) + [h - gd/(t + t_0)^2] \ln[(t + t_0)/t_{now} \quad (5\text{-A}.10)$$

At t = 0

$$a(0) = (t_0/t_{now})^{g[1 + d/t_0]} \quad (5\text{-A}.11)$$

$$H(0) = [g + gd/t_0]/t_0 + [h - gd/t_0^2] \ln(t_0/t_{now}) \quad (5\text{-A}.12)$$

Note a(0) and H(0) are not zero, thus avoiding a Big Bang catastrophe.
For large t

$$a(t_{now}) = 1 \qquad \text{Normalization} \quad (5\text{-A}.13)$$

$$a(t) \cong (t/t_{now})^{g + h(t + t_0)} \quad (5\text{-A}.14)$$

as in eq. 22.1a assuming d and t_0 are small with d/t << 1 and t >> t_0. For large t

$$H(t) = g/t + h(1 + h \ln(t/t_{now})) \quad (5\text{-A}.15)$$

as in eq. 22.2 above. Eq. 5-A.15 follows if d = 0 is substituted in eq. 5-A.10.

5-A.5 A Universe Quantum Vacuum Polarization Basis for the New Model
We now proceed to determine the constants d and t_0 since g and h are determined by the large time behavior of H(t). The equations fixing d and t_0 are

$$a(0) \cong (t_0/t_{now})^{g[1 + d/t_0]} = a_{BBRW}(0) \cong 3.19 \times 10^{-93} \quad (5\text{-A}.16)$$

$$H(0) = [g + gd/t_0]/t_0 + [h - gd/t_0^2] \ln(t_0/t_{now}) = H_{BBRW}(0,0) \cong 3.58 \times 10^{218} \text{ km s}^{-1} \text{ Mpc}^{-1}$$
$$= 1.1 \times 10^{199} \text{ s}^{-1} \quad (5\text{-A}.17)$$

For small $t_0 < 10^{-100}$ s we find H(0) is well approximated by

$$H(0) \cong - (gd/t_0^2) \ln(t_0/t_{now}) \quad (5\text{-A}.18)$$

up to a few per cent. Taking the logarithm of a(0) we see

$$\ln(a(0))/H(0) = - t_0(t_0/d + 1) \quad (5\text{-A}.19)$$

From

$$\ln(a(0)) = g(1 + d/t_0) \ln(t_0/t_{now}) \quad (5\text{-A}.20)$$

we see

$$d = t_0 [\ln(a(0))/\ln(t_0/t_{now}) - g]/g \quad (5\text{-A}.21)$$

As a result

$$\ln(a(0))/H(0) = - t_0(1 + 1 /((1/g)(\ln a(0)/\ln(t_0/t_{NOW})) - 1)) \quad (5\text{-A}.22)$$

which fixes t_0 to be

$$t_0 = 1.936 \times 10^{-197} \text{ s} \qquad (5\text{-A}.23)$$

and

$$d = 2.956 \times 10^{-194} \text{ s} \qquad (5\text{-A}.24)$$

by eq. 5-A.21. Eq. 5-A.23 numerically is

$$\ln(a(0))/H(0) \cong -t_0 \qquad (5\text{-A}.25)$$

to good approximation, which implies

$$a(0) = \exp(-t_0 H(0)) \qquad (5\text{-A}.26)$$

a form similar to quantum mechanical expressions. In the Big Bang region ($t < t^{-200}$ s) $a(t)$ and $H(t)$ are constant by Fig. 5-A.2 so

$$a(t) = \exp[-(t + t_0)H(t)] \qquad (5\text{-A}.27)$$

for $t < t^{-200}$ s since $t_0 = 1.936 \times 10^{-197}$ s dominates in eq. 5-A.27.

5-A.6 A Possible of the Hubble Parameter as a Hamiltonian in the Big Bang Region

The Hubble Parameter is viewed as a time dependent variable. Eq. 5-A.27 raises the possibility that it might be treated as a time dependent Hamiltonian, if appropriately formulated, with eq. 5-A.27 generalized to

$$a(t) = \exp[-(t + t_0)H(t)] \qquad (5\text{-A}.28)$$

for "all" time. Then $a(t)$ becomes a kind of wave function, while retaining its role as the scale factor of the universe. This possibility, which is evocative of our QED-like Model, remains to be explored in detail.

5-A.7 Consequences of the New Blaha Model

Fig. 5-A.1 shows the general form of our model for the $a(t)$ scale factor from $t = 0$ to the present. Figs. 5-A.2 - 5-A.7 show a plot of the Hubble Parameter of eq. 5-A.10 from $t = 0$ to the present.

The model has an impressive set of features:

1. The scale factor becomes very small, but non-zero, as $t \rightarrow 0$, the Big Bang point. The scale factor $a(t)$ and $H(t)$ are constant in the Big Bang region from $t = 0$ to 10^{-200} s. The essential singularity part with the parameter d greatly affects a and H in the Big Bang region.

2. H becomes very large as $t \rightarrow 0$.

3. The large time features of a and H are consistent with known data.

4. From t =o to t = 10^{-200} s a(t) and H(t) are flat. This region is the Big Bang region.

5. The constants g and h are determined from H in the large t region. The parameters d and t_0 are determined by a(t), H and g in the Big Bang region.

6. The essential singularity at t = -t_0 *may* represent the point of fermion-antifermion annihilation to create the universe or Megaverse instance.

Thus we have a satisfactory set of parameters for the new scale factor model extending to the Big Bang period.[35] The additional factor with an essential singularity in time yields a satisfactory model for t = 0 to the present.

5-A.7.1 A Big Dip in H(t)

Fig. 5-A.5 shows a Big Dip that we have seen before in earlier work. This feature is to be expected since H(t) declines from a large positive value at small times and rises at large times near the present. *A minimum must be present by simple algebra.*

The Big Dip events took place at:

Big Dip minimum (H $\cong$ -410) at t = t = 4.1199×10^{14} sec.

Since the changeover from a radiation-dominated phase to a matter-dominated phase is approximately estimated to be at:

Radiation – Matter Domination Transition: t $\cong$ 1.48×10^{12} sec.

Due to uncertainties it may coincide with the Big Dip.

It seems reasonable to conclude the transition from radiation-dominated to matter-dominated causes the Big Dip to occur. The matter-dominated phase transition causes shrinkage as shown in a(t) in Fig. 5-A.4. *The universe contracts by one-third!* [36] We attribute the time delay between the transition and the Big Dip in a(t) to the time required for the transition to occur. (The universe is large at that time after all)

5-A.7.2 Universe Contraction – Early Massive Galaxies

The contraction would appear to "squeeze" the mass-energy in the universe giving it a "belly" (the Big Belly??? of "squeezed" mass-energy).This mass-energy contraction leads to the early formation of galaxies that disperse due to gravitation in the 13.5 Gyrs that follow. The subsequent expansion would also appear to create a "wake" similar to the wake of a water wave.

---

[35] The values of a(0) and H(0) in eq. 5-A.16 and 5-A.17 appear in our Big Bang model. Other values in other physically reasonable models will lead to similar results.
[36] Rather like the condensation of water vapor to liquid.

Evidence[37] has been found for the existence of a huge population of very massive galaxies (39+ have been found so far) that were created within one billion years after the Big Bang. This population of early galaxies is inconsistent with the standard present-day models of galaxy formation. The Big Dip occurs at 2.76 million years – well before one billion years – consistent with the formation of early massive galaxies.

A concentration of mass-energy due to the contraction of the universe appears to present a possible solution. Universe contraction was not considered in the creation of models of galaxy formation.

Another possible source of universe concentration (and voids) of energy appears in our Quantum Big Bang Model. The cause is a large difference in expansion rates (Hubble Constant variations) at the center of the Big Bang compared to the outer edge of the Big Bang.

5-A.7.3 Overshoot in H(t)

The result of the Radiation-Matter transition seems to be a negative H(t) for the energy density Ω_T. H(t) "overshoots" and becomes negative. Crudely put, the clumping of matter in the matter-dominated phase appears to introduce a compactness that results in a decrease in universe size and concentrations of energy.

5-A.7.4 Voids and Bubbles in Space after the Big Dip

The Big Dip concentrates mass-energy at the contraction. The subsequent expansion creates a "type" of wave that generates massive galaxies (bubbles of mass-energy), and also voids – bubbles of space devoid of galaxies. In the course of the following thirteen or so billion years gravitation causes a dispersion of galaxies, voids and bubbles leading to the present day observed distribution.

5-A.7.5 Mystery of the Big Dip in H(t) - A Scenario

At the Big Dip H(t) changes from a declining to a rising trajectory. Based on this fact and the Big Bang model presented in Blaha (2019c) the following scenario seems reasonable:

1. The initial peak, and immediate decline, in H(t) is due to the Y black body radiation phase pressure that decreases rapidly after the Big Bang metastate ends. Thus Ω_T declines rapidly with the Y pressure decline. (Note Ω_T is a sum of energy density and pressure.)

2. The peak in Ω_T reflects an influx of energy (from the Megaverse?) that causes H(t) to begin increasing. There is also a dip below zero in H(t) signifying the shrinkage of the universe as a(t).

3. Afterwards Ω_T continues to be significant and increasing as a(t) and H(t) rise to the present time.

---

[37] T. Wang *et al*, Nature **572**, 211 (2019).

4. In the future Ω_T should continue to rise. The energy increase that this situation implies suggests a certain reality to our Big_Bang-Quantum_Vacuum Theory. .

5-A.8 Radius of Big Bang Universe

The current radius of the universe is estimated[38] to be 4.314×10^{28} cm. Using the estimate in eq. 5-A.16 we find the radius of the Big Bang region at t = 0 is the not unreasonable value:

$$r_{\text{Big Bang}} = 1.376 \times 10^{-64} \text{ cm} \qquad (5\text{-A}.29)$$

or

$$r_{\text{Big Bang}} = 8.532 \times 10^{-32}/M_{\text{Planck}}$$

5-A.9 Particles vs. Universes/Megaverses

How does an elementary particle differ from a universe or Megaverse particle? It seems that elementary particles are physically produced on-shell from on-shell particles. Universe and Megaverse Particles are produced by far off-shell, highly energetic initial particles.

5-A.10 A QED-like Vacuum Polarization Model of H(t)

The origin of the original model in a QED-like vacuum polarization and its extension based on similar considerations suggests that universes may well be types of particles as we picture in Octonion Cosmology. The same view may apply to Megaverses. They may also be types of particles.

If the vacuum polarization viewpoint is correct then it would suggest all universes have a similar pattern of growth within the framework of space 6 of Octonion Cosmology. It would also suggest that all Megaverses have a similar pattern of growth within the framework of space 5 of Octonion Cosmology.

---

[38] M. Tanabashi *et al*, Phys. Rev. **D98**, 030001 (2018).

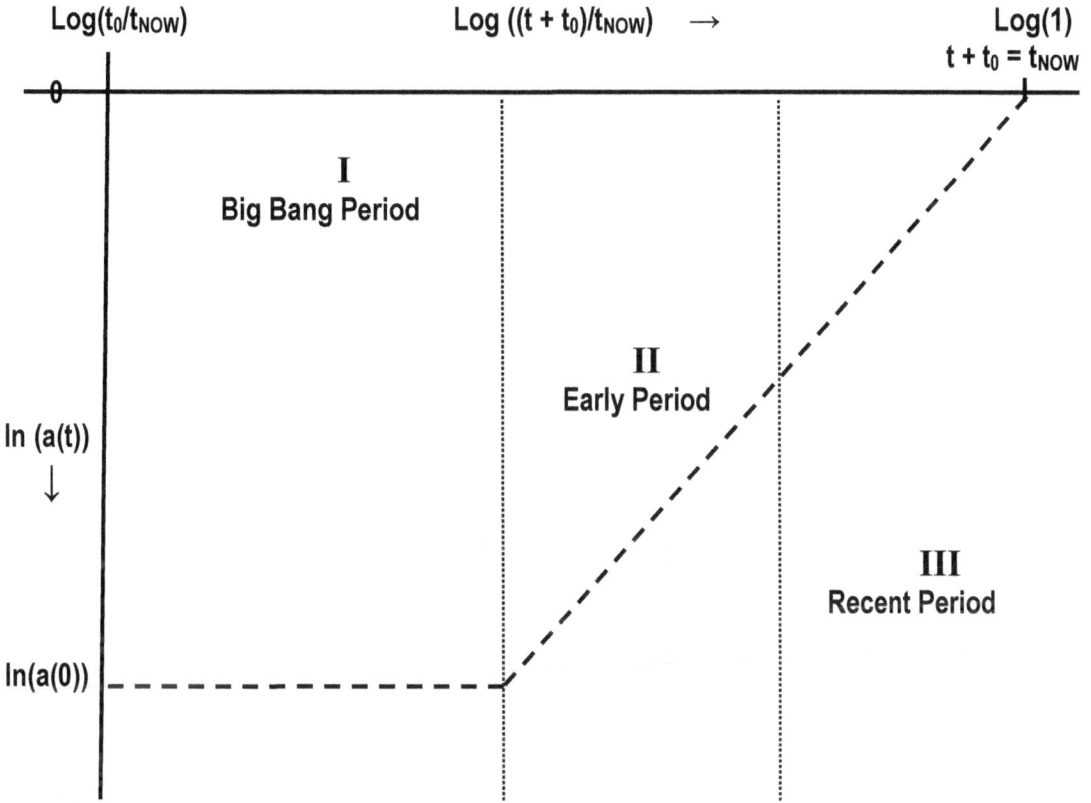

Figure 5-A.1. General form of a(t) of eq. 5-A.9 from t = 0 to t_{NOW} (the present,) It is not a straight line. The three periods are described below. Ln(a(0)) is a constant (and negative) since a(0) is non-zero.

5-A.11 Regions of a(t)

Our vacuum polarization-like model displayed in Fig. 5-A.1 has three regions in time:

I The "Near" Big Bang Period

a(t) in this period is based on the full expression in eq. 5-A.10 with h = 0.

II The Larger time Big Bang Period

a(t) in this period is based on the full expression in eq. 5-A.7 .

III The "Large" time Period

a(t) in this period is based on the full expression in eq. 22.1 above.

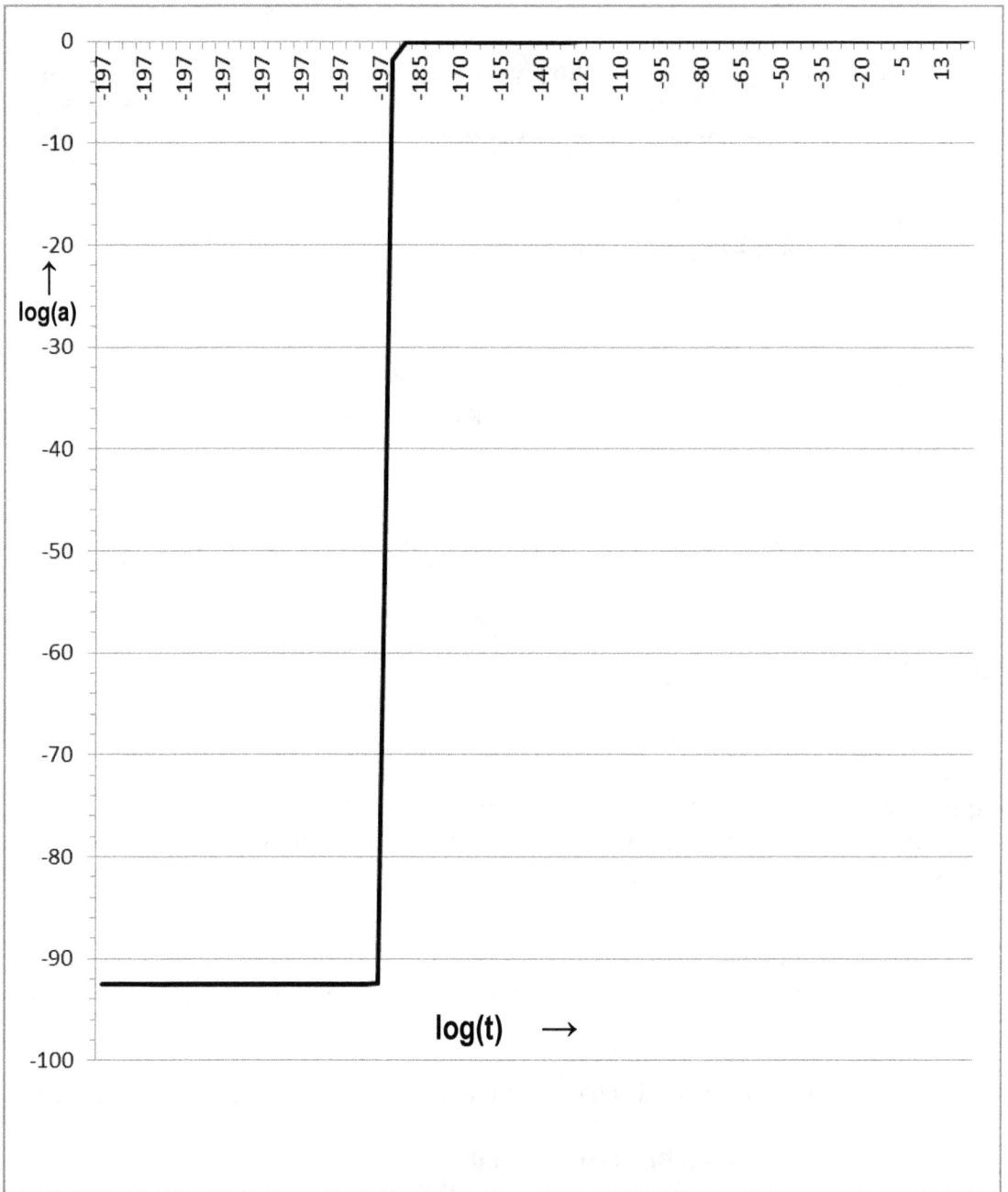

Figure 5-A.2. Plot of $\log_{10}(a(t))$ of eq. 5-A.9 from t = 0 to the present, t_{NOW}. Time (in s) and a(t) are plotted logarithmically to base 10: $\log_{10}(t)$ and $\log_{10}(a)$.. Note a(t) = 3.32495×10^{-93} until t = 10^{-200} s. After a short interval a(t) ranges from 0.87 to 1.

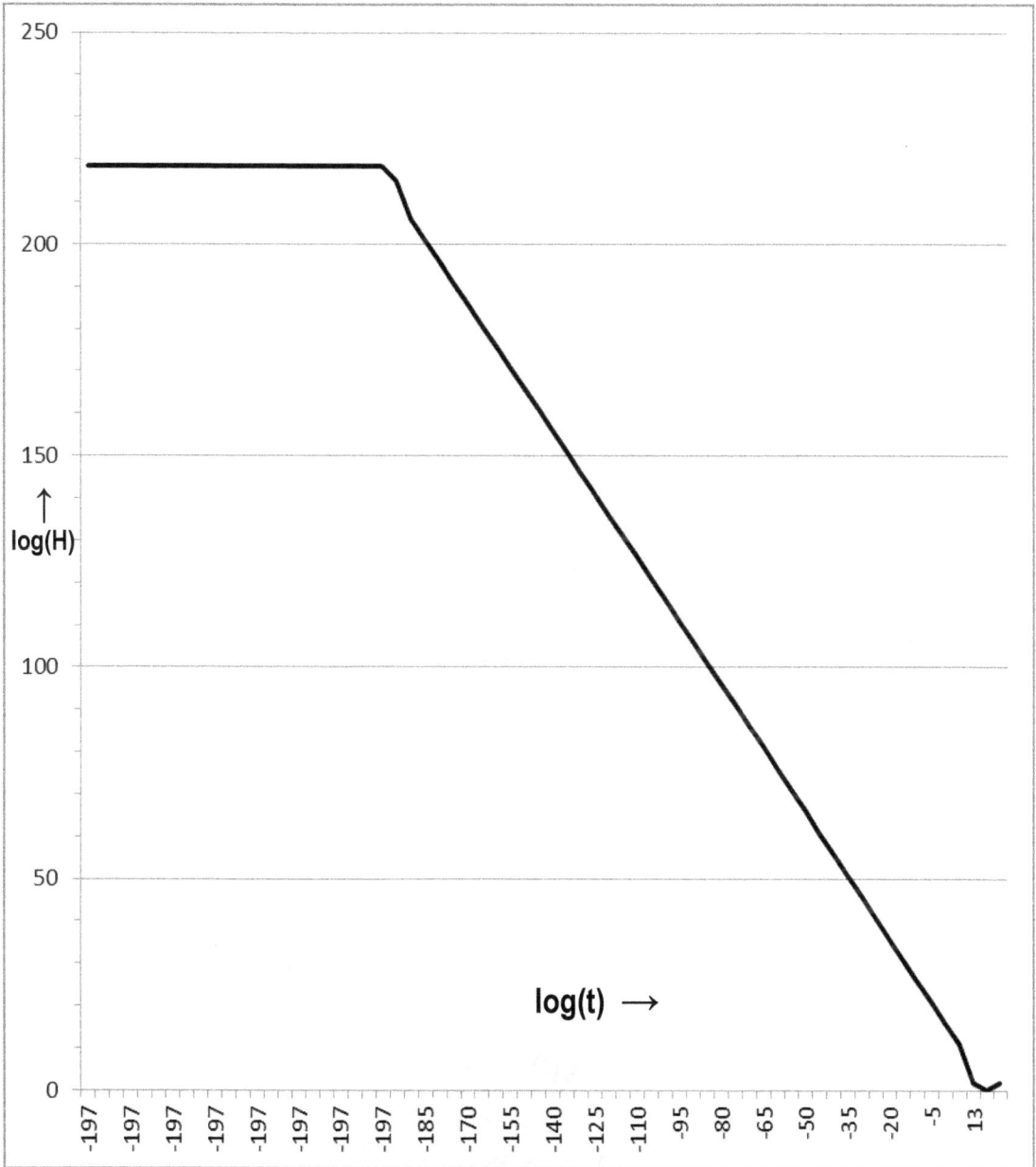

Figure 5-A.3. Plot of the log of the Hubble Parameter $\log_{10}(H(t))$ of eq. 5-A.10 from t = 0 to the present, t_{NOW}. Time (in s) and the Hublle Parameter (in km s^{-1} Mpc^{-1}) are plotted logarithmically to base 10: $\log_{10}(t)$ and $\log_{10}(H)$. Note H(t) = 3.5686×10^{218} km s^{-1} Mpc^{-1} until t = 10^{-200} s. Then H(t) ranges from 1.3×10^{215} km s^{-1} Mpc^{-1} to 68 km s^{-1}Mpc^{-1}.

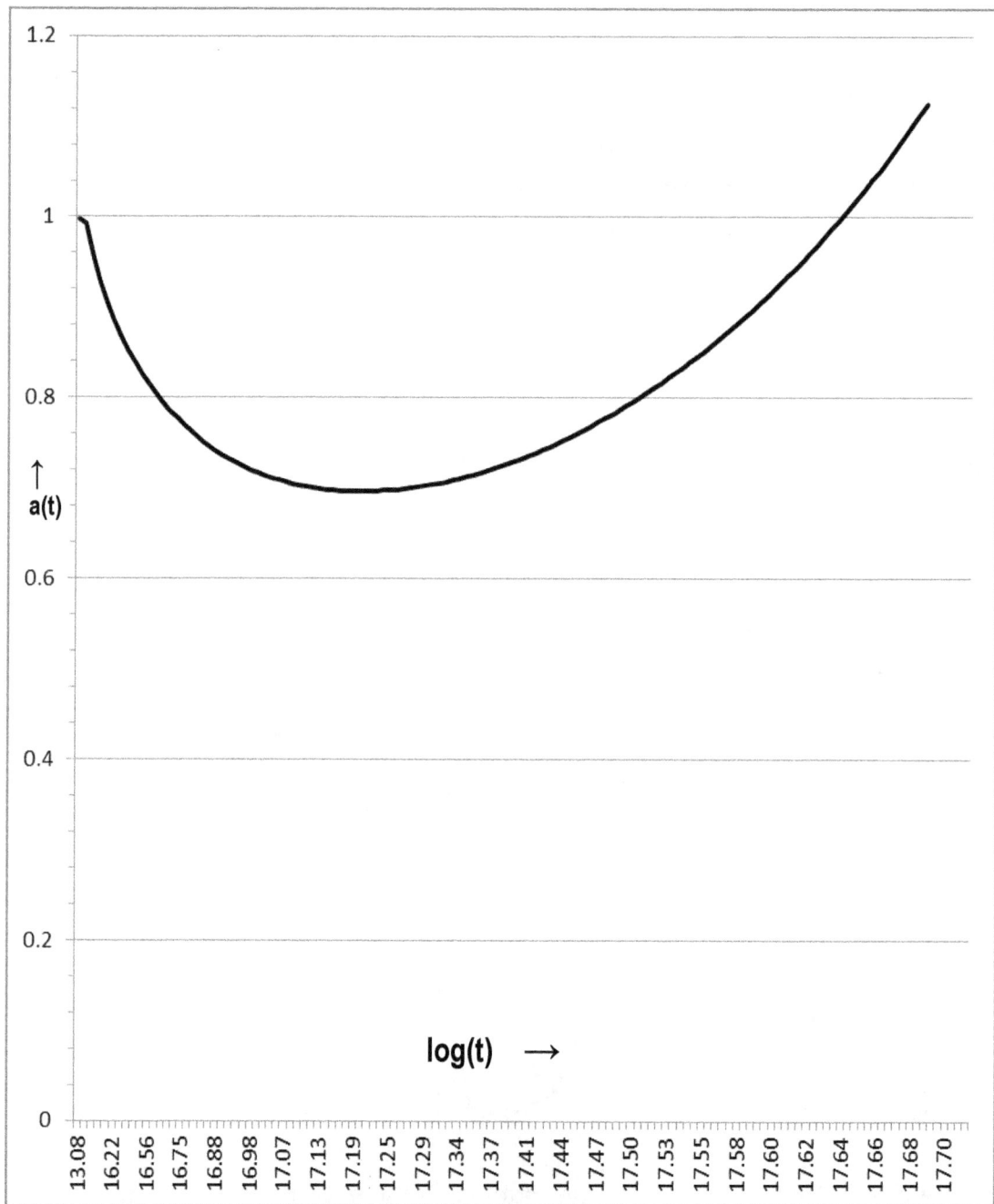

Figure 5-A.4. Plot of a(t) from eq. 5-A.9 vs. $\log_{10}(t)$ from t = $1.198{\times}10^{13}$ s to t = $5.08{\times}10^{17}$ s. Note a(t_{NOW}) = 1. The region of the minimum of a(t) corresponds to the Big Dip region of H(t).

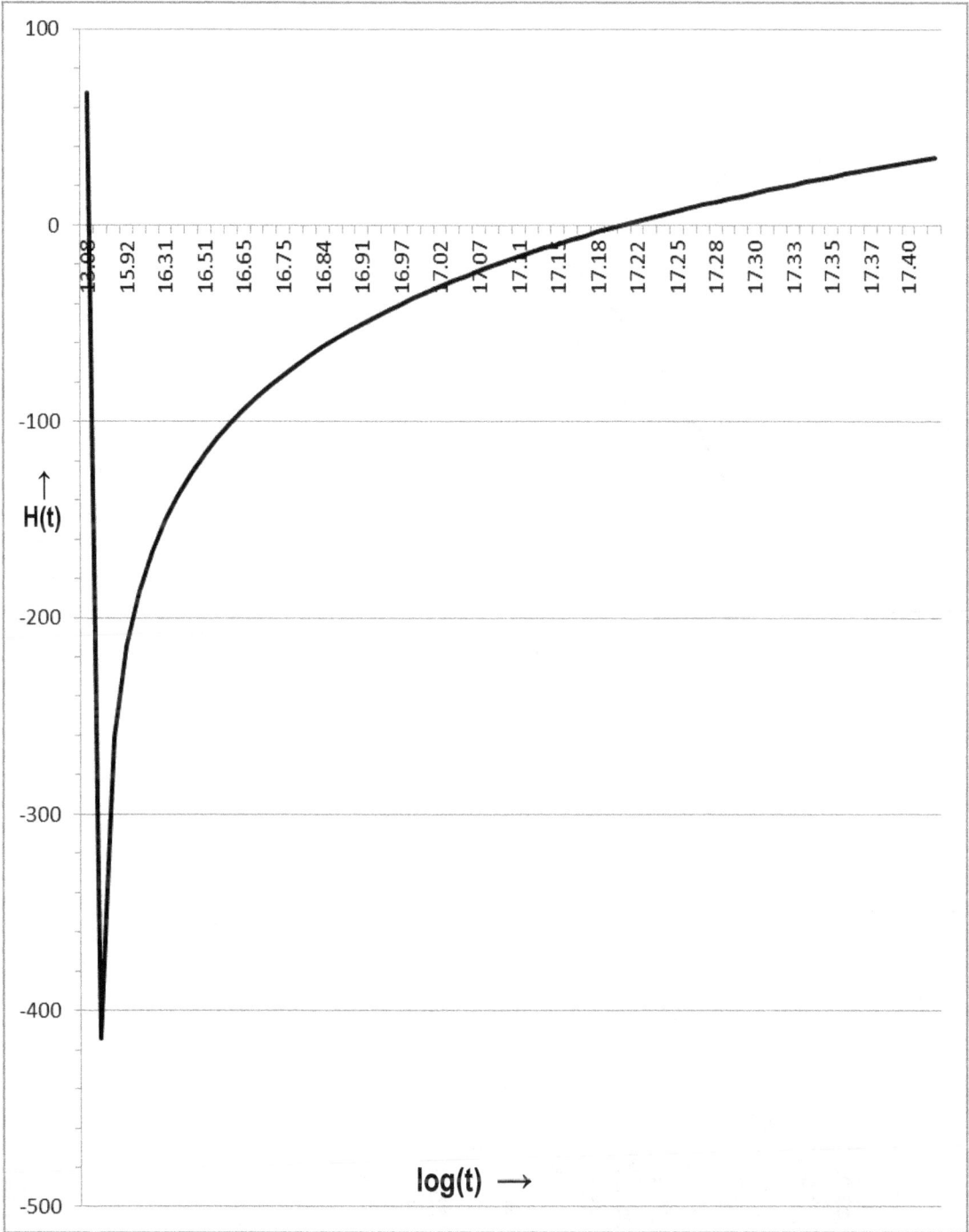

Figure 5-A.5. Plot of H(t) of eq. 5-A.10 vs. $\log_{10}(t)$ from t = 1.198 × 10^{13} to t = 5.08 × 10^{17} s. The Big Dip, the minimum of H(t), occurs at t = 4.1199 × 10^{14} s "shortly" after the radiation–matter transition in the universe.

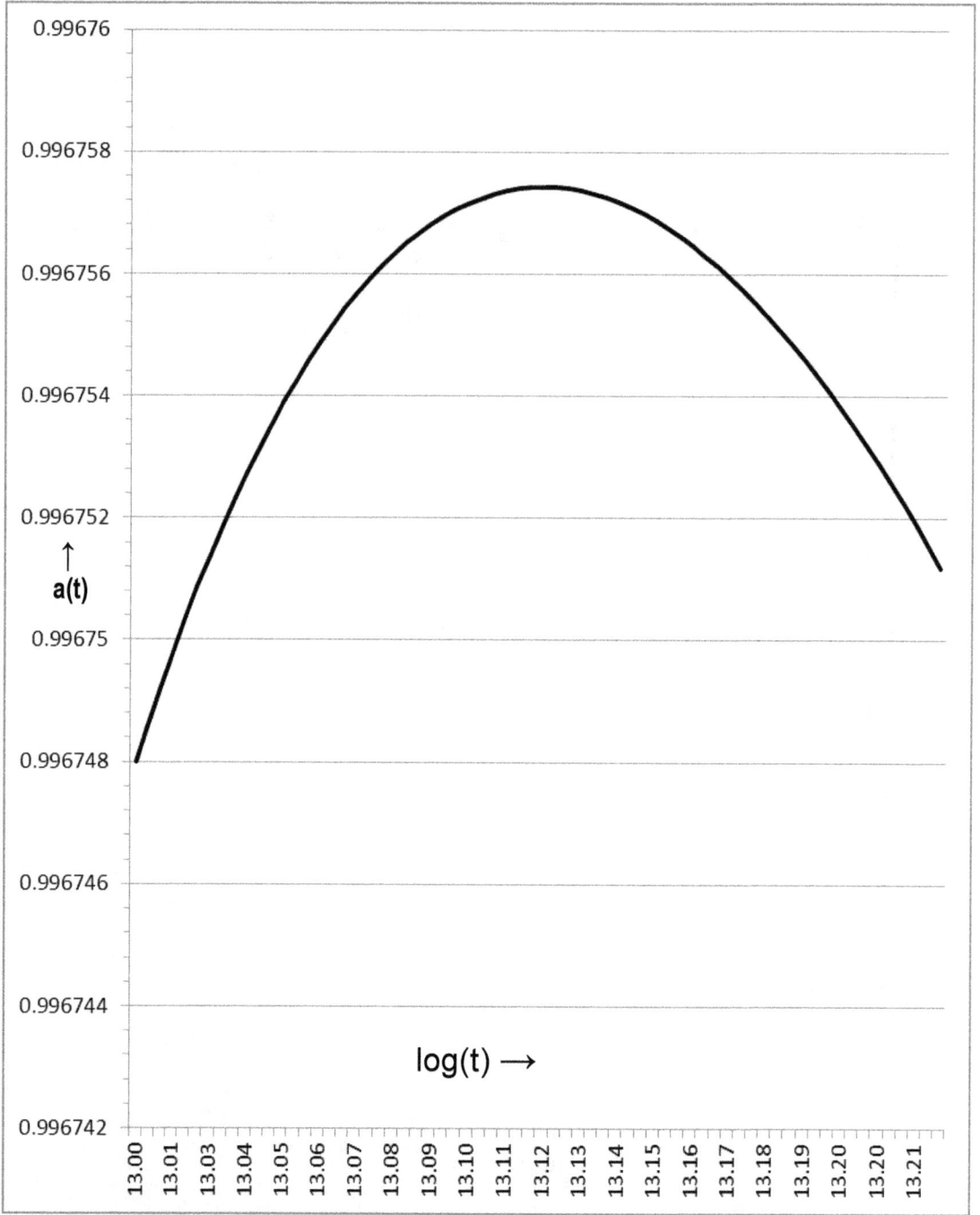

Figure 5-A.6. Plot of a(t) of eq. 5-A.9 vs. $\log_{10}(t)$ around the 380,000 year point from t = 1.01 × 10^{13} to t = 1.65 × 10^{13} s. The 380,000 year point corresponds to $\log_{10}(t)$ = 13.0786.

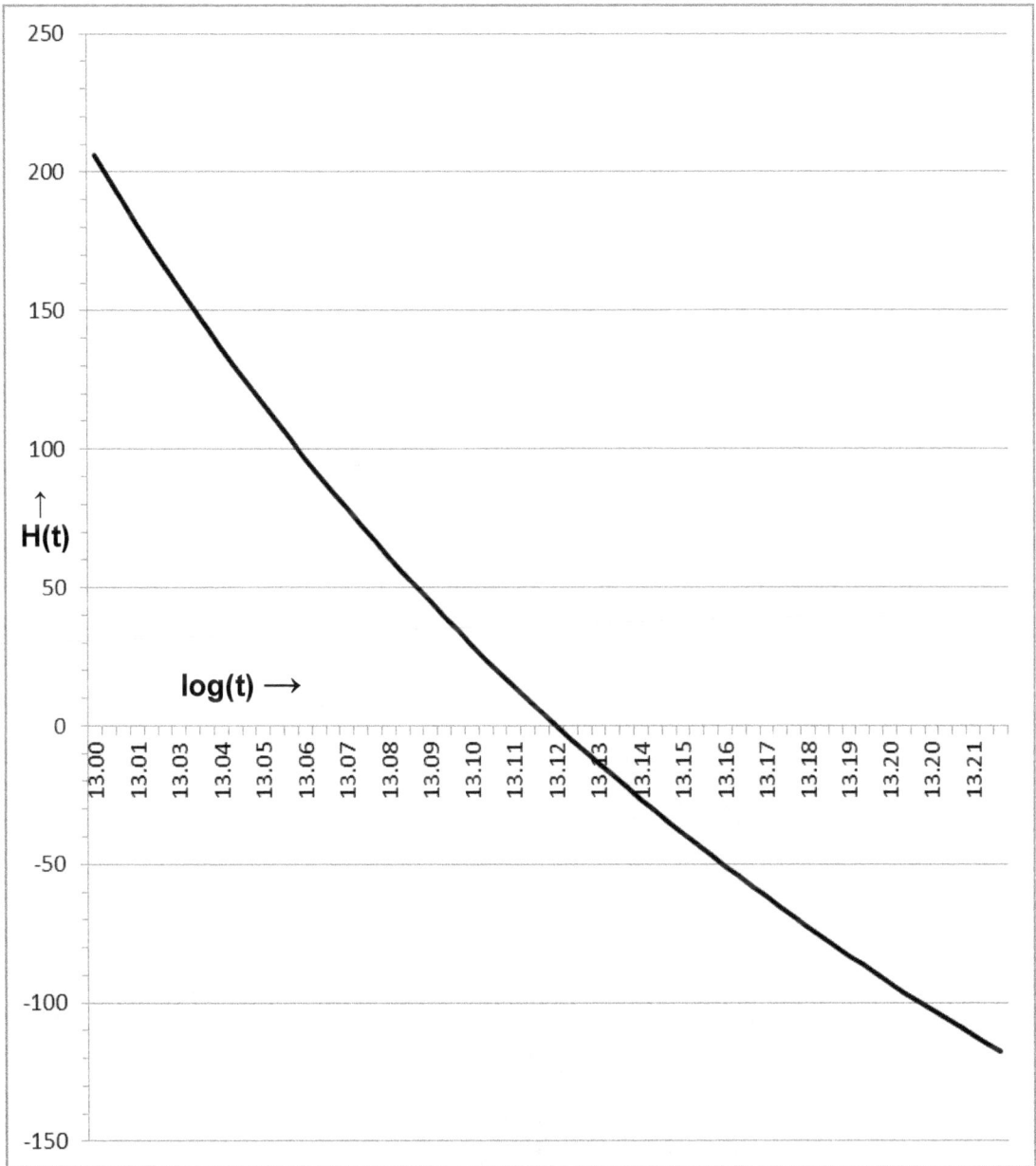

Figure 5-A.7. Plot of H(t) of eq. 5-A.10 vs. $\log_{10}(t)$ around the 380,000 year point from t = 1.01×10^{13} to t = 1.65×10^{13} s. The H(380000 years) value is 67.8 km s^{-1} Mpc^{-1}.

6. Origin of Universes & Megaverses at Essential Singularities

We have presented evidence for the particle-like nature of universes from a comparison of QED vacuum polarization and universe expansion. We saw that the requirement that the size of the universe be finite at the origin of the universe at $t = 0$ led to the introduction of t_0, and the introduction of an essential singularity at the "non-physical" point $t = -t_0$ in analogy with an essential singularity of the QED Fine Structure Eigenvalue function.

An essential singularity at $t = -t_0$ makes physical sense if it reflects the transition of a fermion-antifermion annihilation to a universe particle. We considered the implications of this process in our study of Octonion Cosmology.[39]

The essential singularity that we introduced in eqs. 5.1 and 5.2:

$$a(t) = [(t + t_0)/t_{now}]^{gd/(t + t_0)}[(t + t_0)/t_{now}]^g [(t + t_0)/t_{now}]^{h(t + t_0)} \qquad (5.2)$$
$$\mathbf{I} \qquad\qquad \mathbf{II} \qquad\qquad \mathbf{III}$$

had the form

$$a_I(t) = [(t + t_0)/t_{now}]^{gd/(t + t_0)} \qquad (6.1)$$

The presence of the $(t + t_0)$ terms in eq. 5.2 guaranteed the essential singularity was not at $t = 0$, which marks the beginning of the Big Bang period. Rather it appeared at the "unphysical" time $t = = -t_0$. The interval from $t == -t_0$ to $t = 0$ may be viewed as the generation time of the beginning of the Big Bang period. This time interval must be extremely short. We tentatively used the value:

$$t_0 = 1.936 \times 10^{-197} \text{ s} \qquad (5\text{-A.}23)$$

6.1 Importance of the Essential Singularity

The essential singularity is important because it "zeroes" the impact of the colliding fermion-antifermion state on the scale factor $a(t)$ by trivializing the boundary value conditions to:

$$d^n a_I(t)/dt^n = 0 \qquad (6.2)$$

Vestiges of the fermion-antifermion state are brought to zero by the boundary value conditions. The other factors in $a(t)$ (II and III) have a negligible impact for small t.

---

[39] See Blaha (2021c).

6.2 Plots of H(t) and a(t) Near the Essential Singularity

Figs. 6.1 and 6.2 show plots of H(t) and a(t) near $t = t_0$. As expected $H(t) \to +\infty$ and $a(t) \to 0$.

The Big Bang state appears at $t = 0$ with finite $a(0) \neq 0$ and finite $H(0) < \infty$.

6.3 Megaverse Scale Factor and Hubble Parameter

Megaverses, which are also generated by fermion-antifermion annihilation in Octonio Cosmology, may be expected to have similar features.

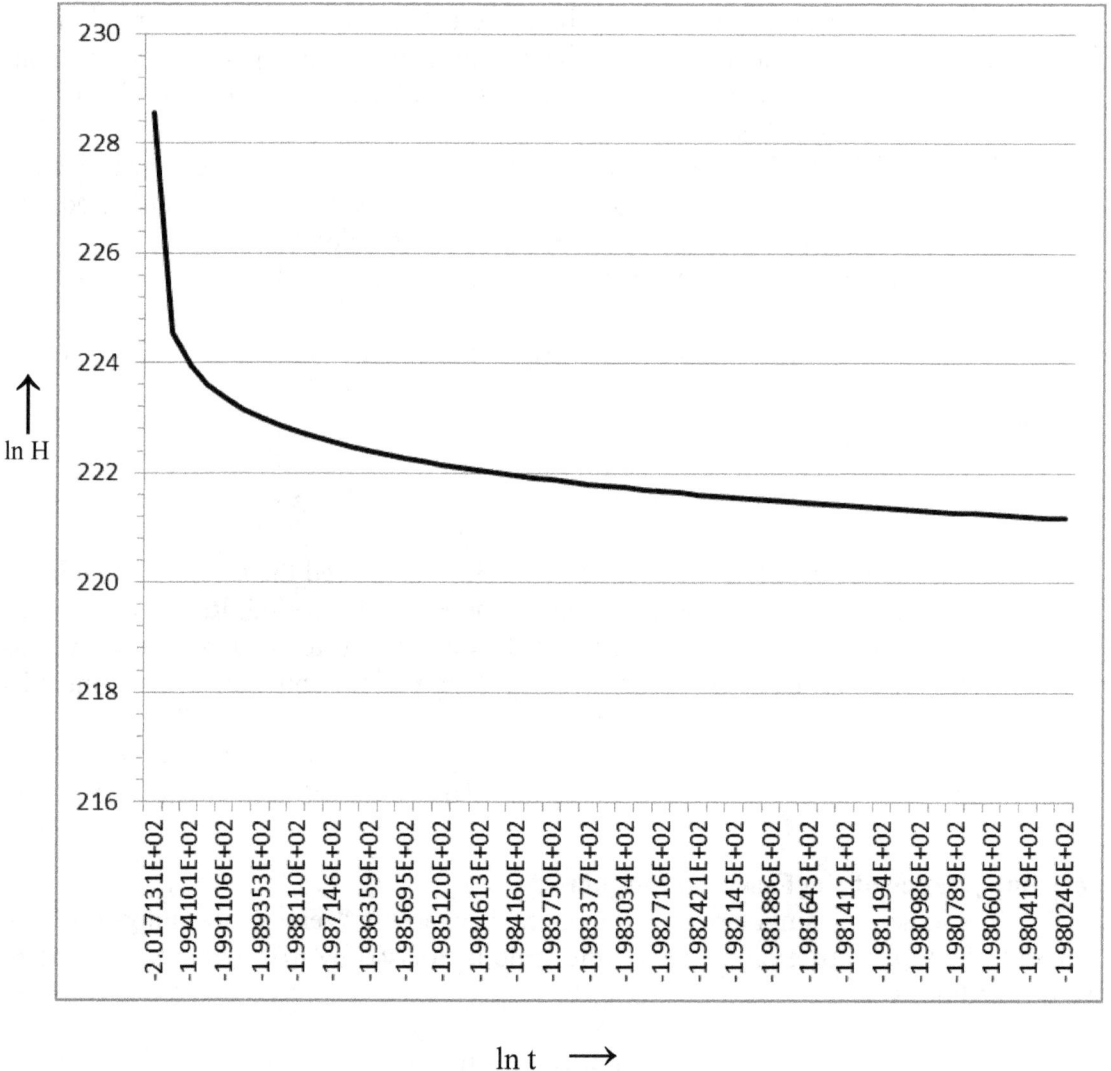

Figure 6.1 A plot of ln H vs. ln t near the essential singularity in a(t). As $t \to t_0$ H(t) goes to +∞.

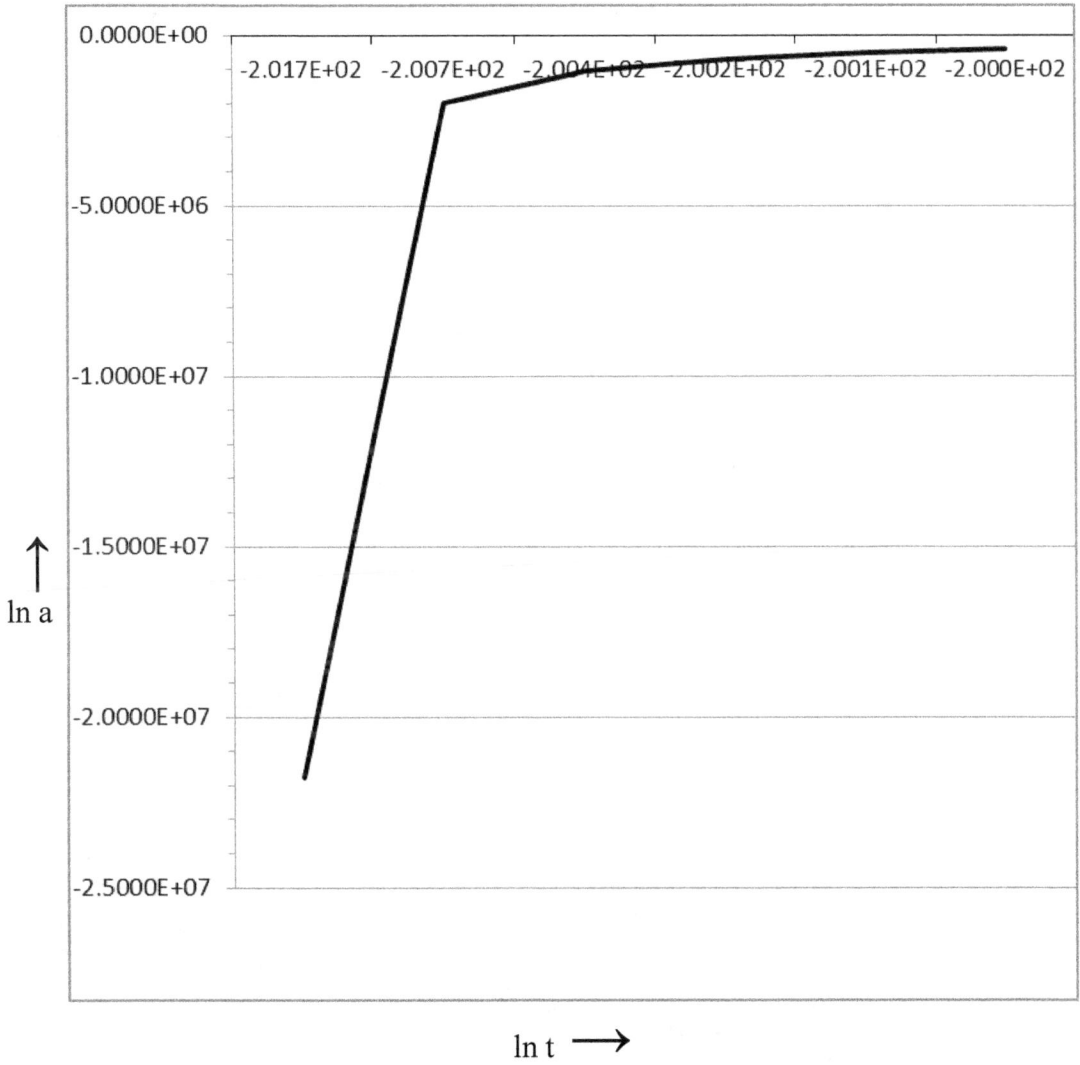

Figure 6.2 A plot of ln a(t) vs. ln t near the essential singularity in a(t). As t → t_0 a(t) goes to zero.

7. QED Vacuum Polarization and Other Standard Model Interactions

The remarkable calculation of the QED Fine Structure Constant as evidenced by Table 2.1 leads us to consider the determination of the coupling constants of the other Standard Model interactions; ElectroWeak SU(2) and Strong SU(3).

In Blaha (2019b) we generalized the F_2 function (and related functions in eq. 1.8) to the cases of the ElectroWeak interaction and the Strong interaction coupling constants by inserting a group theoretic factor in the equation for α_G only:

$$(\alpha_G/2\pi) = c_G^{-1}[gA_4 - (4 + 2g)A_2]/(A_4A_1 - A_2A_3) \qquad (1.13)$$

$$c_G^{-1} = [(11/3)C_{ad} - 2C_f/3]/(16\pi)^3 \qquad (1.14)$$

where C_{ad} is the dimension of the fundamental representation of the group[40] and C_f is the number of fermions (fermion flavor) of the interaction.[41]

The known vector interactions and coupling constants of The Standard Model of Particles are:[42]

- The Strong interaction coupling constant[43] $e_S = 1.22$
- The Weak SU(2) coupling constant $e_W = 0.619$
- The Electromagnetic U(1) coupling constant $e_{QED} = 0.303$

We found good approximations to the SU(2) and SU(3) coupling constants in Blaha (2019b).

The gauge interaction coupling constants, which we denoted with the label G with $e_G = (4\pi\alpha_G)^{1/2}$, for QED, Weak SU(2) and Strong SU(3) have a remarkable regularity—they double from interaction to interaction:[44]

The deeper significance of this regularity is not known.[45]

---

[40] See W. E. Caswell, Phys. Rev. Lett. **33**, 244 (1974) and references therein.

[41] See Blaha (2019b) for the SU(2) and SU(3) values of c_G and the plots of all Standard Model eigenvalue functions.

[42] All coupling constant values are based on data extracted from C. Patrignani *et al* (Particle Data Group), Chinese Physics **C40**, 100001 (2014).

[43] Based on the running coupling constant value $\alpha_s(M_Z^2) = 0.1193 \pm 0.0016$.

[44] Chapter 5 shows the universe scale factor g is ½ of the QED Fine Structure g deepening the mystery.

[45] Given the appearance of factors of two in the Cayley numbers of Octonion Cosmology the factors of two found here may reflect Octonion Cosmology at some deep level.

Group	Known Coupling Constant e_G	Known $e_G^2/(4\pi)$	Calculated $\alpha_G = e_G^2/(4\pi)$	Calculated[46] Exponent g_G
QED, U(1)	0.30282212	$\alpha^{-1} = 137.035999084$	$\alpha^{-1} = 137.035999084$	-0.00058053691948
SU(2)	0.619	0.0305	0.0425	0.54
SU(3)	1.22	0.118	0.086	0.5605

The relative closeness of the calculated values of "fine structure constants" to the experimentally known values is very encouraging—particularly in the case of the Electromagnetic fine structure constant α. It puts to rest other possible explanations for its value.

Our QED calculation of α has no free (adjustable) parameters unlike other attempts in the past. It also is totally based on Quantum Field Theory. The calculation of the non-abelian coupling constants also has no free (adjustable) parameters.

Thus the coupling constant eigenfunctions depend only on inherent perturbation theory based on dynamics. Coupling constant values cannot be "tweaked" to their known values by adjusting input parameters.

The ability of our 1973 calculation of the JBW eigenvalue function together with the new insights into understanding of the precise method to obtain its "fine structure constant" eigenvalues is also encouraging. It opens the possibility that The Standard Model has within itself the mechanism for determining the constants appearing within it. It raises the hope that a similar self-determination mechanism may also exist within the theory to determine the masses appearing in the Higgs particles sector of the theory.

7.1 The Same Standard Model of Particles in All Universes?

Since we saw earlier that 4D universe(s) have the same evolutionary pattern of expansion and contraction, and since The Standard Model[47] coupling constants (which govern all macroscopic and chemical interactions) are determined internally within quantum field theory we can assert that the Physics, Chemistry, and Biology of all 4D universes are the same. And the prospects of life are also the same!

7.2 Doubling of Coupling Constants

The universe vector interaction which we denote with the quantum field label Y with $e_U = (4\pi\alpha_U)^{1/2}$, and the other coupling constants for QED, ElectroWeak SU(2) and Strong SU(3) have a remarkable regularity—they double from interaction to interaction

---

[46] They appear in eqs. 5.5 – 5.8 in Blaha (2019b). See Blaha (2019b) for more details.

[47] One could suggest that The Standard Models of other universes are different. However the simplicity of the group structure would argue otherwise—as would consideration of the case of colliding universes.

as Fig. 7.1 shows. It is interesting that Cayley numbers in Octonion Cosmology also display doubling. The relation between these doubling phenomena, if any, merits further investigation..

INTERACTION	COUPLING CONSTANT[48]
Y Interaction e_U	0.152
QED $e_{QED} = (4\pi\alpha_{QED})^{1/2}$	0.303
Weak SU(2) g_W	0.619
Strong SU(3) g_S	1. 22

Figure 7.1. The interaction constants show a regular doubling. The cause of the doubling is not apparent.

---

[48] M. Tanabashi *et al* (Particle Data Group), Phys. Rev. D**98**, 030001 (2018).

8. Possible Evidence for the Existence of a Megaverse

This chapter provides supporting experimental and theoretical evidence for universe particles.

8.1 Theoretical and Experimental Support

Why are we not content with one universe given its enormous size and variety? It appears that there are important theoretical reasons, and some important experimental observations, that suggest that there is more than our universe 'out there.'

In this chapter[49] we will discuss theoretical reasons and experimental suggestions of a larger space—that we call the *Megaverse*—that contains our universe and, most likely, other universes. The existence of a Megaverse resolves several theoretical issues and may address some important astronomical puzzles that have appeared in recent years.

The theoretical issues, which have been subjects of discussion for many years, are:

1. The need for a 'clock' to measure 'time' knowing that it is to some extent relative and local.
2. The need for a 'quantum observer' to complete the understanding of quantum gravity as described by the Wheeler-DeWitt equation and in other efforts to develop a quantum gravity.
3. The need for other universes to provide theoretical measuring platforms for quantities beyond the charge and mass of the universe. We think here of the other quantum numbers of particles and particle number operators such as Baryon number.
4. The need for an ultimate source of mass and inertia in our universe.

In Blaha (2015a) and earlier books we have suggested that there are weighty reasons to believe that other universes exist.[50] The existence of other universes is a solution to these problems.

These problems have a source in Quantum Gravity and the interpretation of the Wheeler-DeWitt equation in particular. We now consider the issues raised above.

8.1.1 Universe Clocks

Asynchronous Logic provides the equivalent of a clock for the synchronization of processes within large electrical systems such as VLSI chips. Similarly there is a

---

[49] Most of this chapter appears in Blaha (2015a) and in earlier books by the author.

[50] In Blaha (2013a), before the Higgs particle was discovered at CERN we suggested an alternate mechanism was possible if a sister universe existed (making the existence of other universes a reasonable possibility. The Higgs discovery makes the sister universe mechanism unlikely.

need for a universal clock for our universe. As DeWitt[51] points out in his studies of quantum gravity,

'"The variables … [of the quantized Friedmann model] because of their lack of hermiticity, are not rigorously observable and hence cannot yield a measure of proper time which is valid under all circumstances. … . It is for this reason that we may say that "time" is only a phenomenological concept … If the principle of general covariance is truly valid then the quantum mechanics of everyday usage with its dependence on the Schrödinger equations … is only a phenomenological theory. For the only "time" which a covariant theory can admit is an intrinsic time defined by the contents of the universe itself. Any intrinsically defined time is necessarily non-Hermitean, which is equivalent to saying that there exists no clock, whether geometrical or material, which can yield a measure of time which is operationally valid under *all* circumstances, and hence there exists no operational method for determining the Schrödinger state function with arbitrarily high precision."

The lack of a clock within our universe invalidates quantum mechanics in principle and Quantum Gravity in particular. DeWitt concludes, "Thus [quantum gravity] will say nothing about time unless a clock to measure time is provided."

Unruh[52] also has an issue with the source of time:

"One of the key problems is that of time. We see and experience the world in terms of time. We see things grow, develop, and change. However, time does not enter into the Euclidean formulation of quantum gravity directly. In the usual Hamiltonian formulation, the Hamiltonian for quantum gravity is made up of densities which are the generators, not only of spatial coordinate transformations, but also of temporal coordinate transformations. The content of four of Einstein's equations is that some generators are zero. Thus all wave functions are invariant under all spatial and all temporal coordinate transformations. There is nothing in the wave function or the amplitudes which refers to the coordinate t, or the corresponding points of the manifold in any way. How then do we recover the indubitable and ubiquitous experience we have of time? The standard answer is that our experience of time is actually an experience of different correlations between physical quantities in the world. Time is replaced by the readings of clocks. I know that time has changed, not through any direct experience with time, but because the hands of my watch have changed.

Although the implementation of this idea is actually extremely difficult in practice, and although I personally believe that one should formulate one's quantum theory of gravity so as to contain time explicitly, let us nevertheless pursue the consequences of this idea of time as defined internally, as the "reading" of a dynamic variable. For an observer inside the theory, his "time" is not the coordinate t. Rather his

[51] DeWitt, B. S., Phys. Rev. **160**, 1113 (1987).
[52] Unruh, W. G., Phys. Rev. D **40**, 1053 (1989).

time is some one of the given dynamic variables of the theory: y or P. Thus although the coupling to the baby universes via the effective action S is independent of the coordinates t or x, that does not mean that the observer inside the theory will experience the interactions as being independent of time. For him and/or her, time is one of the dynamic variables and so it can depend on the various dynamic variables of the theory, even if it does not depend on the time coordinate t. In general one would expect the observer to see what looks to him like a time-dependent interaction with the baby universes. At one time, some one of the baby universes may couple strongly to the large universe, while at some other time, another of the baby universes will couple more strongly."

In Blaha (2015a) and earlier books, we suggested the existence of other universes provides a 'clock' in principle for our universe. And being universes, these other universes are excellent clocks. DeWitt points out,

"Because every clock has a "one-sided" energy spectrum, its ultimate accuracy must necessarily be inversely proportional to its rest mass. When the whole universe is cast in the role of a clock, the concept of time can of course be made fantastically accurate (at least in principle) … "

Setting a mass scale using other universes, also sets[53] a time scale and resolves the issue of a clock for our universe. *In principle the existence of other universes validates the role of time in the Copenhagen interpretation of Quantum Mechanics.*

8.1.2 Quantum Observer

Attempts to create a quantum gravity theory have to confront the need for an O*bserver* in any quantum theory within the context of the Copenhagen interpretation. DeWitt points out,

"The Copenhagen view depends on the assumed a priori existence of a classical level to which all questions of observation may ultimately be referred. Here, however, the whole universe is the object of inspection; there is no classical vantage point, and hence the interpretation question must be re-argued from the beginning. While we do not wish to stress this point unduly, since, after all, the Friedmann model ignores the vast complexities of the real universe, it is nevertheless clear that the quantum theory of space-time must ultimately force a deviation from the traditional Copenhagen doctrine." And Unruh states

"One of the key features in the interpretation of such transition amplitudes, or wave functions, is the idea that we, as observers are also a part of the Universe as a whole. We, as physical observers, must be describable from within the theory and not as observers external to the theory as in usual quantum mechanics. In usual quantum

---

[53] For example the Planck time value is set by the Planck mass.

mechanics, the interpretation is usually given in terms of observers that are outside of the theory. There one makes a split, with the quantum world at one side of the split, and the observer on the other. von Neumann argued that the predictions of quantum mechanics, at least under certain assumptions, are independent of the exact location of that split, but Bohr argued adamantly for the necessity of such a split (classical observers and quantum world). *There is a great difficulty in setting up such a split for physical observers contained within and influenced by a quantum universe,* [italics added] and for the Universe as a whole, especially including gravity, one cannot argue that the predictions will be independent of where one puts the split. Since all energies interact gravitationally, and our observations are surely energetic phenomenon, the treatment of the energetics of observation as classical would lead to different predictions than if they were treated quantum mechanically. One is therefore forced to devise an interpretation of quantum mechanics in which the observer is part of the quantum system, rather than outside the quantum system.

This means that the interpretation of these transition amplitudes becomes somewhat non-intuitive. One must ask what the system looks like from within, from the viewpoint of an observer who is part of that world, rather than being able to interpret them directly in terms of probabilities for observations made by an external observer."

While the O*bserver* question is addressed by a number of authors, the proposed answers are not entirely convincing. *The existence of other universes provides macroscopic Quantum Observers for our universe.* And our universe provides a macroscopic quantum observer for other universes. Thus the quantum observer issue is resolved.

These considerations lead us to view the existence of other universes as a critical solution to the above problems.

8.1.3 The Higgs Mechanism is Explainable by Extra Dimensions

The Higgs Mechanism 'explains' (generates) fermion and boson masses. However the Higgs potential contains a quadratic term with a constant with the dimensions of [mass]. In a sense the Higgs Mechanism trades one mass for another. From where do the Higgs potentials' masses come?

A further explanation is needed is to determine the origin of the "dimensionful" mass terms in the Higgs' particle equations themselves. At present little if any thought has been given to the origin of these terms. We suggested that, excluding a *deus ex machina* source, the only known way to generate these mass terms in the Higgs' equations is through the separation of equations technique of differential equations. This technique requires additional parameters which can only be the coordinates of *extra unknown dimensions*. The best example of the generation of mass terms appears in the Schwarzschild solution of General Relativity where a separation constant, often denoted M, appears that has the dimension of [mass].

Thus extra space-time dimensions would resolve the origin of Higgs potentials' masses. Given extra dimensions it is reasonable to expect that these extra dimensions contain universes. Thus the Megaverse!

8.1.4 Possible Accretion of Megaverse Matter to Fuel Expansion of Our Universe

If matter is distributed outside of universes in the Megaverse, and if this matter can be accreted to universes by gravitational attraction, then the apparent increasing expansion of our universe may be due to this accretion. In chapter 14 of Blaha (2017c) we presented a model in which this possibility is realized. If true, then we would have tangible evidence of the residence of our universe in the Megaverse.

8.1.5 Asynchronous Logic is a Requirement of Universes

By establishing Asynchronous Logic principles[54] as the basis for the existence of universes and for setting the number of dimensions in each universe – four; and basis of fermion particles - qubes – we have found deeper principles of organization for the foundations of physics. The principles built on this foundation serve to enable the coordination of complex physical processes.

Usually we look at particle processes primarily from a space-time perspective: particles collide and produce new particles. We primarily think of the incoming and outgoing particles in a collision. However, considering the set of fundamental particles – and the particle transforming interactions in themselves – neglecting space-time and momentum considerations – leads us to view particles as constituting an alphabet and their interactions as a type of computer grammar.[55] Then the Asynchronicity Principles enable us to bring in space-time in a way that gives us the maximum complexity with the most minimal assumptions. As Leibniz[56] points out our universe has maximal complexity with minimal assumptions.

8.1.6 The Meaning of Total Quantities of a Universe

The 'external' properties of a universe are normally questioned—for the simple reason that it is assumed that there is no 'outside' of our universe. For example, Misner (1973) asserts:[57]

'There is no such thing as "the energy (or angular momentum, or charge) of a closed universe," according to general relativity, and this for a simple reason. To weigh something one needs a platform on which to stand to do the weighing.'

Misner et al presumes no such platform exists. If there is but one closed universe as most currently believe then one cannot measure any totals of a closed universe (which ours may be to be). Yet if we take a more general view that our universe is only one of

---

[54] The basis of this section is described in detail in Blaha (2015a). That book places Physics within a logical framework that is a possible deeper ground for fundamental Physics theory.

[55] This conceptual approach was first described in Blaha (1998) who went on to characterize our universe as one enormous word evolving in time.

[56] See Rescher (1967).

[57] Pp. 457 - 458.

many then it becomes possible to measure total mass, charge, angular momentum, baryon number, and many other quantities of interest. Indeed, the existence of other universes (within the encompassing Megaverse) opens the door to an understanding of time, mass, energy, and all the other quantities necessary to develop a dynamical theory of universes.

Later we will also see that one can then treat universes as 'particles', and develop 'universe dynamics', which might explain knotty problems such as the Big Bang and its precursor (if any). We will do this in subsequent chapters after first considering the possible structure of universes in general in the Megaverse.

8.2 Possible Experimental Evidence for the Megaverse

At first glance it would seem impossible to produce evidence for the existence of other universes. However there are subtle means by which we can 'sense' experimentally 'nearby' universes should they exist. The mechanism would appear to be gravitational effects exerted on objects within our universe by unseen objects of enormous mass. Currently there appears to be three experimental suggestions of the existence of 'nearby' universes and one theoretical argument based on an influx of mass-energy from the Megaverse that may cause the expansion of our universe.

8.2.1 Great Attractors

One potential support is the discovery of the Great Attractor (at the center of the Laniakea Galaxy Supercluster), and the more massive Shapley Attractor (centered in the Shapley Supercluster)[58]. These attractors contain massive numbers of galaxies and are drawing galaxies over a distance of millions of light years towards them.

If another universe(s) is 'near' our universe it could act as a 'gravitational magnet' and draw galaxies within our universe towards it to form one or more superclusters which could then act as attractors. Thus attractors might indirectly reveal the presence of other nearby universes—contrary to the expected large scale uniformity of the universe. The only other apparent source of superclusters is chance. Chance seems an unsatisfactory possibility in the present case.

8.2.2 Bright Bumps in Universe Suggesting Collision with Another Universe

A recent study[59] of the residual brightness of parts of the accessible universe found that bright patches appeared if a model of the CMB (Cosmic Microwave Background) with gases, stars and dust was 'subtracted' from the PLANCK map of the entire sky. After the subtraction one would expect only noise spread throughout the sky. However, bright patches were seen in a certain range of frequencies. These anomalies are thought to be a result of our universe colliding with another object – presumably another universe in the Megaverse.

---

[58] Tully, R. Brent; Courtois, Helene; Hoffman, Yehuda; Pomarède, Daniel, "The Laniakea Supercluster of galaxies". Nature (4 September 2014). 513 (7516): 71–73; arXiv:1409.0880.
[59] Ranga-Ram Chary, arXiv.org:/1510.00126 (2015).

8.2.3 Cold Spot in Universe Suggesting Collision with Another Universe

Another recent study[60] of a huge cold region of the universe spanning billions of light years revealed that this region is not a relatively empty region but rather is similar to in its distribution of galaxies to the rest of the universe. Previous the Cold Spot (an area where cosmic microwave background radiation – the leftover Big Bang radiation is weak – making it significantly colder (0.00015C colder) than the average temperature of the universe.)

An analysis of 7,000 galaxy redshifts using new high-resolution data has now shown that the Cold Spot is similar to the rest of the universe. The Durham University group suggested that the Cold Spot might have been caused by a collision between our universe and another Universe. They further suggested that there is only a 1 in 50 chance that it could explain by standard cosmology. could produce this feature

Thus we have another important piece of circumstantial evidence in favor of other universes and thus the Megaverse.

8.2.4 Megaverse Energy-Matter Infusion into Our Universe

In chapter 14 of Blaha (2017c) we presented a model for an influx of mass-energy from the Megaverse to support the Bond-Gold-Hoyle-Narlikar Steady State Cosmology, which was originally based on the 'continuous creation of mass-energy' by Hoyle and Narlikar. This model explains why the value of Ω makes the universe close to flat. If this model is correct then we would have concrete support for a Megaverse with a low mass-energy density leaking mass-energy into our universe. *More generally, it suggests that universes are surfaces of high mass-energy density in a Megaverse of low mass-energy density – with a ratio of mass-energy densities of the other of 10^{30}.*

8.2.5 Conclusion

We conclude that data is beginning to emerge favoring multiple universes and a physical Megaverse in support of the theoretical justifications presented earlier.

---

[60] T. Shanks et al, Durham University (Australia), Monthly Notices of the Royal Astronomical Society, 2016 .

REFERENCES

Akhiezer, N. I., Frink, A. H. (tr), 1962, *The Calculus of Variations* (Blaisdell Publishing, New York, 1962).

Bjorken, J. D., Drell, S. D., 1964, *Relativistic Quantum Mechanics* (McGraw-Hill, New York, 1965).

Bjorken, J. D., Drell, S. D., 1965, *Relativistic Quantum Fields* (McGraw-Hill, New York, 1965).

Blaha, S., 1995, *C++ for Professional Programming* (International Thomson Publishing, Boston, 1995).

_____, 1998, *Cosmos and Consciousness* (Pingree-Hill Publishing, Auburn, NH, 1998 and 2002).

_____, 2002, *A Finite Unified Quantum Field Theory of the Elementary Particle Standard Model and Quantum Gravity Based on New Quantum Dimensions™ & a New Paradigm in the Calculus of Variations* (Pingree-Hill Publishing, Auburn, NH, 2002).

_____, 2004, *Quantum Big Bang Cosmology: Complex Space-time General Relativity, Quantum Coordinates,™Dodecahedral Universe, Inflation, and New Spin 0, ½, 1 & 2 Tachyons & Imagyons* (Pingree-Hill Publishing, Auburn, NH, 2004).

_____, 2005a, *Quantum Theory of the Third Kind: A New Type of Divergence-free Quantum Field Theory Supporting a Unified Standard Model of Elementary Particles and Quantum Gravity based on a New Method in the Calculus of Variations* (Pingree-Hill Publishing, Auburn, NH, 2005).

_____, 2005b, *The Metatheory of Physics Theories, and the Theory of Everything as a Quantum Computer Language* (Pingree-Hill Publishing, Auburn, NH, 2005).

_____, 2005c, *The Equivalence of Elementary Particle Theories and Computer Languages: Quantum Computers, Turing Machines, Standard Model, Superstring Theory, and a Proof that Gödel's Theorem Implies Nature Must Be Quantum* (Pingree-Hill Publishing, Auburn, NH, 2005).

_____, 2006a, *The Foundation of the Forces of Nature* (Pingree-Hill Publishing, Auburn, NH, 2006).

_____, 2006b, *A Derivation of ElectroWeak Theory based on an Extension of Special Relativity; Black Hole Tachyons; & Tachyons of Any Spin.* (Pingree-Hill Publishing, Auburn, NH, 2006).

_____, 2007a, *Physics Beyond the Light Barrier: The Source of Parity Violation, Tachyons, and A Derivation of Standard Model Features* (Pingree-Hill Publishing, Auburn, NH, 2007).

_____, 2007b, *The Origin of the Standard Model: The Genesis of Four Quark and Lepton Species, Parity Violation, the ElectroWeak Sector, Color SU(3), Three Visible Generations of Fermions, and One Generation of Dark Matter with Dark Energy* (Pingree-Hill Publishing, Auburn, NH, 2007).

_____, 2008a, A Direct Derivation of the Form of the Standard Model From GL(16) (Pingree-Hill Publishing, Auburn, NH, 2008).

_____, 2008b, *A Complete Derivation of the Form of the Standard Model With a New Method to Generate Particle Masses Second Edition* (Pingree-Hill Publishing, Auburn, NH, 2008)

_____, 2009, *The Algebra of Thought & Reality: The Mathematical Basis for Plato's Theory of Ideas, and Reality Extended to Include A Priori Observers and Space-Time Second Edition* (Pingree-Hill Publishing, Auburn, NH, 2009).

_____, 2010a, *Operator Metaphysics: A New Metaphysics Based on a New Operator Logic and a New Quantum Operator Logic that Lead to a Mathematical Basis for Plato's Theory of Ideas and Reality* (Pingree-Hill Publishing, Auburn, NH, 2010).

_____, 2010b, *The Standard Model's Form Derived from Operator Logic, Superluminal Transformations and GL(16)* (Pingree-Hill Publishing, Auburn, NH, 2010).

_____, 2010c, *SuperCivilizations: Civilizations as Superorganisms* (McMann-Fisher Publishing, Auburn, NH, 2010).

_____, 2011a, *21st Century Natural Philosophy Of Ultimate Physical Reality* (McMann-Fisher Publishing, Auburn, NH, 2011).

_____, 2011b, *All the Universe! Faster Than Light Tachyon Quark Starships & Particle Accelerators with the LHC as a Prototype Starship Drive Scientific Edition* (Pingree-Hill Publishing, Auburn, NH, 2011).

_____, 2011c, *From Asynchronous Logic to The Standard Model to Superflight to the Stars* (Blaha Research, Auburn, NH, 2011).

_____, 2012a, *From Asynchronous Logic to The Standard Model to Superflight to the Stars volume 2: Superluminal CP and CPT, U(4) Complex General Relativity and The Standard Model, Complex Vierbein General Relativity, Kinetic Theory, Thermodynamics* (Blaha Research, Auburn, NH, 2012).

_____, 2012b, *Standard Model Symmetries, And Four And Sixteen Dimension Complex Relativity; The Origin Of Higgs Mass Terms* (Blaha Reasearch, Auburn, NH, 2012).

_____, 2013a, *Multi-Stage Space Guns, Micro-Pulse Nuclear Rockets, and Faster-Than-Light Quark-Gluon Ion Drive Starships* (Blaha Research, Auburn, NH, 2013).

_____, 2013b, *The Bridge to Dark Matter; A New Sister Universe; Dark Energy; Inflatons; Quantum Big Bang; Superluminal Physics; An Extended Standard Model Based on Geometry* (Blaha Reasearch, Auburn, NH, 2013).

_____, 2014a, *Universes and Megaverses: From a New Standard Model to a Physical Megaverse; The Big Bang; Our Sister Universe's Wormhole; Origin of the Cosmological Constant, Spatial Asymmetry of the Universe, and its Web of Galaxies; A Baryonic Field between Universes and Particles; Megaverse Extended Wheeler-DeWitt Equation* (Blaha Reasearch, Auburn, NH, 2014).

_____, 2014b, *All the Megaverse! Starships Exploring the Endless Universes of the Cosmos Using the Baryonic Force* (Blaha Research, Auburn, NH, 2014).

_____, 2014c, *All the Megaverse! II Between Megaverse Universes: Quantum Entanglement Explained by the Megaverse Coherent Baryonic Radiation Devices – PHASERs Neutron Star Megaverse Slingshot Dynamics Spiritual and UFO Events, and the Megaverse Microscopic Entry into the Megaverse* (Blaha Research, Auburn, NH, 2014).

_____, 2015a, *PHYSICS IS LOGIC PAINTED ON THE VOID: Origin of Bare Masses and The Standard Model in Logic, U(4) Origin of the Generations, Normal and Dark Baryonic Forces, Dark Matter, Dark Energy, The Big Bang, Complex General Relativity, A Megaverse of Universe Particles* (Blaha Research, Auburn, NH, 2015).

_____, 2015b, *PHYSICS IS LOGIC Part II: The Theory of Everything, The Megaverse Theory of Everything, U(4)⊗U(4) Grand Unified Theory (GUT), Inertial Mass = Gravitational Mass, Unified Extended Standard Model and a New Complex General Relativity with Higgs Particles, Generation Group Higgs Particles* (Blaha Research, Auburn, NH, 2015).

_____, 2015c, *The Origin of Higgs ("God") Particles and the Higgs Mechanism: Physics is Logic III, Beyond Higgs – A Revamped Theory With a Local Arrow of Time, The Theory of Everything Enhanced, Why Inertial Frames are Special, Universes of the Mind* (Blaha Research, Auburn, NH, 2015).

_____, 2015d, *The Origin of the Eight Coupling Constants of The Theory of Everything: U(8) Grand Unified Theory of Everything (GUTE), S^8 Coupling Constant Symmetry, Space-Time Dependent Coupling Constants, Big Bang Vacuum Coupling Constants, Physics is Logic IV* (Blaha Research, Auburn, NH, 2015).

_____, 2016a, *New Types of Dark Matter, Big Bang Equipartition, and A New U(4) Symmetry in the Theory of Everything: Equipartition Principle for Fermions, Matter is 83.33% Dark, Penetrating the Veil of the Big Bang, Explicit QFT Quark Confinement and Charmonium, Physics is Logic V* (Blaha Research, Auburn, NH, 2016).

72 **REFERENCES**

_____, 2016b, *The Periodic Table of the 192 Quarks and Leptons in The Theory of Everything: The U(4) Layer Group, Physics is Logic VI* (Blaha Research, Auburn, NH, 2016).

_____, 2016c, *New Boson Quantum Field Theory, Dark Matter Dynamics, Dark Matter Fermion Layer Mixing, Genesis of Higgs Particles, New Layer Higgs Masses, Higgs Coupling Constants, Non-Abelian Higgs Gauge Fields, Physics is Logic VII* (Blaha Research, Auburn, NH, 2016).

_____, 2016d, *Unification of the Strong Interactions and Gravitation: Quark Confinement Linked to Modified Short-Distance Gravity; Physics is Logic VIII* (Blaha Research, Auburn, NH, 2016).

_____, 2016e, *MoND: Unification of the Strong Interactions and Gravitation II, Quark Confinement Linked to Large-Scale Gravity, Physics is Logic IX* (Blaha Research, Auburn, NH, 2016).

_____, 2016f, *CQ Mechanics: A Unification of Quantum & Classical Mechanics, Quantum/Semi-Classical Entanglement, Quantum/Classical Path Integrals, Quantum/Classical Chaos* (Blaha Research, Auburn, NH, 2016).

_____, 2016g, *GEMS: Unified Gravity, ElectroMagnetic and Strong Interactions: Manifest Quark Confinement, A Solution for the Proton Spin Puzzle, Modified Gravity on the Galactic Scale* (Pingree Hill Publishing, Auburn, NH, 2016).

_____, 2016h, *Unification of the Seven Boson Interactions based on the Riemann-Christoffel Curvature Tensor* (Pingree Hill Publishing, Auburn, NH, 2016).

_____, 2017a, *Unification of the Eleven Boson Interactions based on 'Rotations of Interactions'* (Pingree Hill Publishing, Auburn, NH, 2017).

_____, 2017b, *The Origin of Fermions and Bosons, and Their Unification* (Pingree Hill Publishing, Auburn, NH, 2017).

_____, 2017c, *Megaverse: The Universe of Universes* (Pingree Hill Publishing, Auburn, NH, 2017).

_____, 2017d, *SuperSymmetry and the Unified SuperStandard Model* (Pingree Hill Publishing, Auburn, NH, 2017).

_____, 2017e, *From Qubits to the Unified SuperStandard Model with Embedded SuperStrings: A Derivation* (Pingree Hill Publishing, Auburn, NH, 2017).

_____, 2017f, *The Unified SuperStandard Model in Our Universe and the Megaverse: Quarks, ... ,* (Pingree Hill Publishing, Auburn, NH, 2017).

_____, 2018a, *The Unified SuperStandard Model and the Megaverse SECOND EDITION A Deeper Theory based on a New Particle Functional Space that Explicates Quantum Entanglement Spookiness (Volume 1)* (Pingree Hill Publishing, Auburn, NH, 2018).

_____, 2018b, *Cosmos Creation: The Unified SuperStandard Model, Volume 2, SECOND EDITION* (Pingree Hill Publishing, Auburn, NH, 2018).

_____, 2018c, *God Theory (*Pingree Hill Publishing, Auburn, NH, 2018).

_____, 2018d, *Immortal Eye: God Theory: Second Edition* (Pingree Hill Publishing, Auburn, NH, 2018).

_____, 2018e, *Unification of God Theory and Unified SuperStandard Model THIRD EDITION* (Pingree Hill Publishing, Auburn, NH, 2018).

_____, 2019a, *Calculation of: QED α = 1/137, and Other Coupling Constants of the Unified SuperStandard Theory* (Pingree Hill Publishing, Auburn, NH, 2019).

_____, 2019b, *Coupling Constants of the Unified SuperStandard Theory SECOND EDITION* (Pingree Hill Publishing, Auburn, NH, 2019).

_____, 2019c, *New Hybrid Quantum Big_Bang–Megaverse_Driven Universe with a Finite Big Bang and an Increasing Hubble Constant* (Pingree Hill Publishing, Auburn, NH, 2019).

_____, 2019d, *The Universe, The Electron and The Vacuum* (Pingree Hill Publishing, Auburn, NH, 2019).

_____, 2019e, *Quantum Big Bang – Quantum Vacuum Universes (Particles)* (Pingree Hill Publishing, Auburn, NH, 2019).

_____, 2019f, *The Exact QED Calculation of the Fine Structure Constant Implies ALL 4D Universes have the Same Physics/Life Prospects* (Pingree Hill Publishing, Auburn, NH, 2019).

_____, 2019g, *Unified SuperStandard Theory and the SuperUniverse Model: The Foundation of Science* (Pingree Hill Publishing, Auburn, NH, 2019).

_____, 2020a, *Quaternion Unified SuperStandard Theory (The QUeST) and Megaverse Octonion SuperStandard Theory (MOST)* (Pingree Hill Publishing, Auburn, NH, 2020).

_____, 2020b, *United Universes Quaternion Universe - Octonion Megaverse* (Pingree Hill Publishing, Auburn, NH, 2020).

_____, 2020c, *Unified SuperStandard Theories for Quaternion Universes & The Octonion Megaverse* (Pingree Hill Publishing, Auburn, NH, 2020).

_____, 2020d, *The Essence of Eternity: Quaternion & Octonion SuperStandard Theories* (Pingree Hill Publishing, Auburn, NH, 2020).

_____, 2020e, *The Essence of Eternity II* (Pingree Hill Publishing, Auburn, NH, 2020).

_____, 2020f, *A Very Conscious Universe* (Pingree Hill Publishing, Auburn, NH, 2020).

_____, 2020g, *Hypercomplex Universe* (Pingree Hill Publishing, Auburn, NH, 2020).

_____, 2020h, *Beneath the Quaternion Universe* (Pingree Hill Publishing, Auburn, NH, 2020).

_____, 2020i, *Why is the Universe Real? From Quaternion & Octonion to Real Coordinates* (Pingree Hill Publishing, Auburn, NH, 2020).

_____, 2020j, *The Origin of Universes: of Quaternion Unified SuperStandard Theory (QUeST); and of the Octonion Megaverse (UTMOST)* (Pingree Hill Publishing, Auburn, NH, 2020).

_____, 2020k, *The Seven Spaces of Creation: Octonion Cosmology* (Pingree Hill Publishing, Auburn, NH, 2020).

_____, 2020l, *From Octonion Cosmology to the Unified SuperStandard Theory of Particles* (Pingree Hill Publishing, Auburn, NH, 2020).

_____, 2021a, *Pioneering the Cosmos* (Pingree Hill Publishing, Auburn, NH, 2021).

_____, 2021b, *Pioneering the Cosmos II* (Pingree Hill Publishing, Auburn, NH, 2021).

_____, 2021c, *Beyond Octonion Cosmology* (Pingree Hill Publishing, Auburn, NH, 2021).

Eddington, A. S., 1952, *The Mathematical Theory of Relativity* (Cambridge University Press, Cambridge, U.K., 1952).

Fant, Karl M., 2005, *Logically Determined Design: Clockless System Design With NULL Convention Logic* (John Wiley and Sons, Hoboken, NJ, 2005).

Feinberg, G. and Shapiro, R., 1980, *Life Beyond Earth: The Intelligent Earthlings Guide to Life in the Universe* (William Morrow and Company, New York, 1980).

Gelfand, I. M., Fomin, S. V., Silverman, R. A. (tr), 2000, *Calculus of Variations* (Dover Publications, Mineola, NY, 2000).

Giaquinta, M., Modica, G., Souchek, J., 1998, *Cartesian Coordinates in the Calculus of Variations* Volumes I and II (Springer-Verlag, New York, 1998).

Giaquinta, M., Hildebrandt, S., 1996, *Calculus of Variations* Volumes I and II (Springer-Verlag, New York, 1996).

Gradshteyn, I. S. and Ryzhik, I. M., 1965, *Table of Integrals, Series, and Products* (Academic Press, New York, 1965).

Heitler, W., 1954, *The Quantum Theory of Radiation* (Claendon Press, Oxford, UK, 1954).

Huang, Kerson, 1992, *Quarks, Leptons & Gauge Fields 2nd Edition* (World Scientific Publishing Company, Singapore, 1992).

Jost, J., Li-Jost, X., 1998, *Calculus of Variations* (Cambridge University Press, New York, 1998).

Kaku, Michio, 1993, *Quantum Field Theory*, (Oxford University Press, New York, 1993).

Kirk, G. S. and Raven, J. E., 1962, *The Presocratic Philosophers* (Cambridge University Press, New York, 1962).

Landau, L. D. and Lifshitz, E. M., 1987, *Fluid Mechanics 2nd Edition*, (Pergamon Press, Elmsford, NY, 1987).

Misner, C. W., Thorne, K. S., and Wheeler, J. A., 1973, *Gravitation* (W. H. Freeman, New York, 1973).

Rescher, N., 1967, *The Philosophy of Leibniz* (Prentice-Hall, Englewood Cliffs, NJ, 1967).

Rieffel, Eleanor and Polak, Wolfgang, 2014, *Quantum Computing* (MIT Press, Cambridge, MA, 2014).

Riesz, Frigyes and Sz.-Nagy, Béla, 1990, *Functional Analysis* (Dover Publications, New York, 1990).

Sagan, H., 1993, *Introduction to the Calculus of Variations* (Dover Publications, Mineola, NY, 1993).

Sakurai, J. J., 1964, *Invariance Principles and Elementary Particles* (Princeton University Press, Princeton, NJ, 1964).

Streater, R. F. and Wightman, A. S., 2000, *PCT, Spin, Statistics, and All That* (Princeton University Press, Princeton, NJ 2000).

Weinberg, S., 1972, *Gravitation and Cosmology* (John Wiley and Sons, New York, 1972).

Weinberg, S., 1995, *The Quantum Theory of Fields Volume I* (Cambridge University Press, New York, 1995).

Weinberg, S., 2000, *The Quantum Theory of Fields Volume III Supersymmetry* (Cambridge University Press, New York, 2000).

Weyl, H., 1950, *Space, Time, Matter* (Dover, New York, 1950).

INDEX

About the Author

Stephen Blaha is a well-known Physicist and Man of Letters with interests in Science, Society and civilization, the Arts, and Technology. He had an Alfred P. Sloan Foundation scholarship in college. He received his Ph.D. in Physics from Rockefeller University. He has served on the faculties of several major universities. He was also a Member of the Technical Staff at Bell Laboratories, a manager at the Boston Globe Newspaper, a Director at Wang Laboratories, and President of Blaha Software Inc. and of Janus Associates Inc. (NH).

Among other achievements he was a co-discoverer of the "r potential" for heavy quark binding developing the first (and still the only demonstrable) non-Aeolian gauge theory with an "r" potential; first suggested the existence of topological structures in superfluid He-3; first proposed Yang-Mills theories would appear in condensed matter phenomena with non-scalar order parameters; first developed a grammar-based formalism for quantum computers and applied it to elementary particle theories; first developed a new form of quantum field theory without divergences (thus solving a major 60 year old problem that enabled a unified theory of the Standard Model and Quantum Gravity without divergences to be developed); first developed a formulation of complex General Relativity based on analytic continuation from real space-time; first developed a generalized non-homogeneous Robertson-Walker metric that enabled a quantum theory of the Big Bang to be developed without singularities at $t = 0$; first generalized Cauchy's theorem and Gauss' theorem to complex, curved multi-dimensional spaces; received Honorable Mention in the Gravity Research Foundation Essay Competition in 1978; first developed a physically acceptable theory of faster-than-light particles; first derived a composition of extremums method in the Calculus of Variations; first quantitatively suggested that inflationary periods in the history of the universe were not needed; first proved Gödel's Theorem implies Nature must be quantum; provided a new alternative to the Higgs Mechanism, and Higgs particles, to generate masses; first showed how to resolve logical paradoxes including Gödel's Undecidability Theorem by developing Operator Logic and Quantum Operator Logic; first developed a quantitative harmonic oscillator-like model of the life cycle, and interactions, of civilizations; first showed how equations describing superorganisms also apply to civilizations. A recent book shows his theory applies successfully to the past 14 years of history and to *new* archaeological data on Andean and Mayan civilizations as well as Early Anatolian and Egyptian civilizations.

He first developed an axiomatic derivation of the form of The Standard Model from geometry – space-time properties – The Unified SuperStandard Model. It unifies all the known forces of Nature. It also has a Dark Matter sector that includes a Dark ElectroWeak sector with Dark doublets and Dark gauge interactions. It uses quantum coordinates to remove infinities that crop up in most

interacting quantum field theories and additionally to remove the infinities that appear in the Big Bang and generate inflationary growth of the universe. It shows gravity has a MOND-like form without sacrificing Newton's Laws. It relates the interactions of the MOND-like sector of gravity with the r-potential of Quark Confinement. The axioms of the theory lead to the question of their origin. We suggest in the preceding edition of this book it can be attributed to an entity with God-like properties. We explore these properties in "God Theory" and show they predict that the Cosmos exists forever although individual universes (or incarnations of our universe) "come and go." Several other important results emerge from God Theory such a functionally triune God. The Unified SuperStandard Theory has many other important parts described in the Current Edition of *The Unified SuperStandard Theory* and expanded in subsequent volumes.

Blaha has had a major impact on a succession of elementary particle theories: his Ph.D. thesis (1970), and papers, showed that quantum field theory calculations to all orders in ladder approximations could not give scaling deep inelastic electron-nucleon scattering. He later showed the eigenvalue equation for the fine structure constant α in Johnson-Baker-Willey QED had a zero at $\alpha = 1$ not 1/137 by solving the Schwinger-Dyson equations to all orders in an approximation that agreed with exact results to 4th order in α thus ending interest in this theory. In 1979 at Prof. Ken Johnson's (MIT) suggestion he calculated the proton-neutron mass difference in the MIT bag model and found the result had the wrong sign reducing interest in the bag model. These results all appear in Physical Review papers. In the 2000's he repeatedly pointed out the shortcomings of SuperString theory and showed that The Standard Model's form could be derived from space-time geometry by an extension of Lorentz transformations to faster than light transformations. This deeper space-time basis greatly increases the possibility that it is part of THE fundamental theory. Recently, Blaha showed that the Weak interactions differed significantly from the Strong, electromagnetic and gravitation interactions in important respects while these interactions had similar features, and suggested that ElectroWeak theory, which is essentially a glued union of the Weak interactions and Electromagnetism, possibly modulo unknown Higgs particle features, be replaced by a unified theory of the other interactions combined with a stand-alone Weak interaction theory. Blaha also showed that, if Charmonium calculations are taken seriously, the Strong interaction coupling constant is only a factor of five larger than the electromagnetic coupling constant, and thus Strong interaction perturbation theory would make sense and yield physically meaningful results.

In graduate school (1965-71) he wrote substantial papers in elementary particles and group theory: The Inelastic E- P Structure Functions in a Gluon Model. Phys. Lett. B40:501-502,1972; Deep-Inelastic E-P Structure Functions In A Ladder Model With Spin 1/2 Nucleons, Phys.Rev. D3:510-523,1971; Continuum Contributions To The Pion Radius, Phys. Rev. 178:2167-2169,1969; Character Analysis of U(N) and SU(N), J. Math. Phys. 10, 2156 (1969); and The Calculation of the Irreducible Characters of the Symmetric Group in Terms of the

Compound Characters, (Published as Blaha's Lemma in D. E. Knuth's book: *The Art of Computer Programming Vols. 1 – 4*).

In the early 1980's Blaha was also a pioneer in the development of UNIX for financial, scientific and Internet applications: benchmarked UNIX versions showing that block size was critical for UNIX performance, developing financial modeling software, starting database benchmarking comparison studies, developing Internet-like UNIX networking (1982) and developing a hybrid shell programming technique (1982) that was a precursor to the PERL programming language. He was also the manager of the AT&T ten-year future products development database. His work helped lead to commercial UNIX on computers such as Sun Micros, IBM AIX minis, and Apple computers.

In the 1980's he pioneered the development of PC Desktop Publishing on laser printers and was nominated for three "Awards for Technical Excellence" in 1987 by PC Magazine for PC software products that he designed and developed.

Recently he has developed a theory of Megaverses – actual universes of which our universe is one – with quantum particle-like properties based on the Wheeler-DeWitt equation of Quantum Gravity. He has developed a theory of a baryonic force, which had been conjectured many years ago, and estimated the strength of the force based on discrepancies in measurements of the gravitational constant G. This force, operative in D-dimensional space, can be used to escape from our universe in "uniships" which are the equivalent of the faster-than-light starships proposed in the author's earlier books. Thus travel to other universes, as well as to other stars is possible.

Blaha also considered the complexified Wheeler-DeWitt equation and showed that its limitation to real-valued coordinates and metrics generated a Cosmological Constant in the Einstein equations.

The author has also recently written a series of books on the serious problems of the United States and their solution as well as a book on the decline of Mankind that will follow from current social and genetic trends in Mankind.

In the past twenty years Dr. Blaha has written over 80 books on a wide range of topics. Some recent major works are: *From Asynchronous Logic to The Standard Model to Superflight to the Stars, All the Universe!, SuperCivilizations: Civilizations as Superorganisms, America's Future: an Islamic Surge, ISIS, al Qaeda, World Epidemics, Ukraine, Russia-China Pact, US Leadership Crisis, The Rises and Falls of Man – Destiny – 3000 AD: New Support for a Superorganism MACRO-THEORY of CIVILIZATIONS From CURRENT WORLD TRENDS and NEW Peruvian, Pre-Mayan, Mayan, Anatolian, and Early Egyptian Data, with a Projection to 3000 AD*, and *Mankind in Decline: Genetic Disasters, Human-Animal Hybrids, Overpopulation, Pollution, Global Warming, Food and Water Shortages, Desertification, Poverty, Rising Violence, Genocide, Epidemics, Wars, Leadership Failure.*

He has taught approximately 4,000 students in undergraduate, graduate, and postgraduate corporate education courses primarily in major universities, and large companies and government agencies.

Recently he developed a quantum theory, The Unified SuperStandard Theory (UST), which describes elementary particles in detail without the difficulties of conventional quantum field theory. He found that the internal symmetries of this theory could be exactly derived from an octonion theory called QUeST. He further found that another octonion theory (UTMOST) describes the Megaverse. It can hold QUeST universes such as our own universe. It has an internal symmetry structure which is a superset of the QUeST internal symmetries.